The Communicating Scientist

Olle Bergman · Sarang Park ·
Joanna Bagniewska

The Communicating Scientist

A Practical Handbook

With illustrations by Dorota Paczesniak

Olle Bergman
Eskilstuna, Sweden

Joanna Bagniewska
University of Oxford
Oxford, UK

Sarang Park
Fudan University Shanghai Medical College
Shanghai, China

ISBN 978-3-031-84172-9   ISBN 978-3-031-84173-6 (eBook)
https://doi.org/10.1007/978-3-031-84173-6

© The Editor(s) (if applicable) and The Author(s), under exclusive license to Springer Nature Switzerland AG 2025

This work is subject to copyright. All rights are solely and exclusively licensed by the Publisher, whether the whole or part of the material is concerned, specifically the rights of translation, reprinting, reuse of illustrations, recitation, broadcasting, reproduction on microfilms or in any other physical way, and transmission or information storage and retrieval, electronic adaptation, computer software, or by similar or dissimilar methodology now known or hereafter developed.
The use of general descriptive names, registered names, trademarks, service marks, etc. in this publication does not imply, even in the absence of a specific statement, that such names are exempt from the relevant protective laws and regulations and therefore free for general use.
The publisher, the authors and the editors are safe to assume that the advice and information in this book are believed to be true and accurate at the date of publication. Neither the publisher nor the authors or the editors give a warranty, expressed or implied, with respect to the material contained herein or for any errors or omissions that may have been made. The publisher remains neutral with regard to jurisdictional claims in published maps and institutional affiliations.

This Springer imprint is published by the registered company Springer Nature Switzerland AG
The registered company address is: Gewerbestrasse 11, 6330 Cham, Switzerland

If disposing of this product, please recycle the paper.

Alles was überhaupt gedacht werden kann, kann klar gedacht werden. Alles, was sich aussprechen lässt, lässt sich klar aussprechen.

Everything that can be thought at all, can be thought clearly. Anything that can be said, can be said clearly.

Ludwig Wittgenstein,
Tractatus Logico-Philosophicus

I dedicate my part of this work to Göran Hagström—storyteller from Småland and vir bonus dicendi peritus.

—*Olle Bergman*

For Olle's family.

—*Sarang Park*

To all scientists trying to find their voice.

—*Joanna Bagniewska*

Foreword: Effective Medicine for the Plague of Our Time

As I am writing this foreword, most of the world is still in the midst of a terrible epidemic. No, I am not referring to COVID-19, but to the concurrent epidemic of ineffective communication. Exposure to this disease vector does not produce robust immunity, quite the opposite. It seems that exposure to bad communication potentiates future bad communication: many of the worst techniques end up being imitated by speakers and writers—including educated professionals committed to serving their peers, the public, or their customers.

We need a cure for this cultural epidemic—if not complete immunization, then at least some effective therapies. Luckily, you are currently holding one of these: Olle Bergman, Sarang Park, and Joanna Bagniewska's contribution, which could be labeled: *For treatment of the effects of poor communication. Take two chapters, and call your intellectual physician in the morning.*

That experts in STEMM[1] fields—e.g. scientists, engineers, technologists, and physicians—have trouble communicating their ideas has become a cliché. Why? Because there is a lot of truth in it! Experience also shows that channeling the knowledge of these groups through professional communicators (press secretaries, journalists, politicians) may produce equally bad or even worse results than STEMM professionals speaking for themselves.

Four Major Reasons for Communication Failure

It seems to me that many previous approaches to improving technical communication have failed for four major reasons:

- *They look at the problem as being a unidirectional transmission of information*—from knowledgeable insider to ignorant outsider—rather than as a sophisticated interaction characterized by multiple feedback loops at different stages in the process.

---

[1] Science, technology, engineering, mathematics, medicine.

- ***They focus on only a few kinds of communicative interactions***, such as scientific papers, keynote presentations or major press releases. Accordingly, they do not discuss the myriad of small-scale communications (emails, pitches, summaries, even LinkedIn profiles) that cumulatively convey much more information than large-scale, formal statements.
- ***They are either blissfully unaware of***, or have reflexively rejected, the vast body of knowledge gained from the long tradition of the study of human persuasion: the discipline of rhetoric.
- Even if they do try to draw from the discipline of rhetoric, ***they adopt the attitude that problems of communication and persuasion can be solved by merely a "return to the Classics."*** Thus, they alienate themselves from the most pressing concerns of their audiences. It is still true that many of the answers we need can be found in Aristotle. However, finding more practical approaches in the rhetoric for understanding and dealing with memes, trolls and other phenomena of newer media requires more: a level of cognitive effort that people would rather apply to research and development in their own disciplines, not for communication purposes.

Communication as a System

Bergman, Park, and Bagniewska avoid all of these problems because they treat communication as a *system*. They therefore recognize that interaction operates on multiple levels—often simultaneously. While doing so, they do not restrict themselves to large-scale statements and official presentations. Instead, they line up a plethora of communication activities along different continua: small to large, informal to formal, simple to complex. They discuss communication within and between technical disciplines, as well as communication between people with expert knowledge and those who don't yet have it.

Additionally, by weaving the insights of the Ancients through their discussions of how people—and in particular, STEMM professionals—communicate today, they apply the knowledge of the centuries to contemporary concerns. The odds that present-day researchers will be able to employ the appropriate classical methods are greatly improved by the ways Bergman, Park and Bagniewska adapt these techniques to fit specific communicative situations.

This combination of relentless practicality and abstract insights sets the book apart from the many others that do not even try to arrive at an effective synthesis of the two seemingly dialectically opposed approaches. Bergman, Park, and Bagniewska never lose sight of their ultimate abstract goal: (1) getting ideas that are inside one person's brain into other people's brains, and (2) then convincing those people of the significance and accuracy of those ideas so that (3) the people receiving the ideas think about them in the same way as the people who transmitted them, thus (4) improving the chances that the receivers agree with the originators about the correct actions to take.

The complexity of that previous sentence illustrates the difficulty of juggling abstract concepts and practical tips—and the necessity of doing so. Without abstraction, it becomes impossible to generalize our knowledge to situations other than those specifically described in the book. But without practical details, it takes too much cognitive effort to hold the full abstractions in our minds. Thus by showing precisely how the abstract approaches can play out in material (and utterly familiar), communicative situations, Bergman, Park, and Bagniewska allow readers to incorporate rhetorical thinking into their own practical communication.

A Strong and Effective Medicine

I began this introduction by comparing the wave of bad communication that currently plagues the world to an actual epidemic. In doing so, I broke one of my own personal rules: do not label as an epidemic anything that does not have an identified, physical disease-vector. But, as I noted above, there really is something epidemic-like in this case: the forms of communication used by people of seemingly high social status tend to be imitated, even when these techniques are ineffective! People adopt the approaches of their intellectual leaders even when they probably cannot remember what those leaders were trying to communicate, and in this way bad communication certainly does act like a virus, propagating from mind to mind and spreading damage throughout communities.

And that is why I characterized this book as a treatment. Even if it cannot eliminate transmission, for those afflicted, *The Communicating Scientist* is a strong and effective medicine.

Michael D. C. Drout
Professor of English
Wheaton College
Norton, MA, USA

Preface: A Book That Illuminates, Instructs and Inspires

About This Book, Its Scope and How to Use It

Welcome to our book! To explain its purpose and scope as straightforwardly as possible, we've picked a method from its pages—namely the "Why, Who, What ..." technique.[2] Here we go!

Why (should you read this book)?
The short answer is that you aspire to be *The Communicating Scientist*, mentioned right there on the cover. The longer answer is that we want to help you and your co-readers—most likely scientists, R&D[3] staff and like-minded people—communicate more effectively to reach your professional and personal goals, and inspire action in others. This book will help you become more target-oriented, structured and convincing during different communication activities.

Who (are you, our reader)?
You are most likely a science person[4] in academia, the startup world, or R&D. For different reasons, you are looking for practical advice on how to solve the communication challenges you meet on the path of academic and professional life—both in scientific peer-to-peer communication and science outreach.

In addition, this book may also prove helpful if you are a student at any level—from high school to Ph.D.—when dealing with communication challenges during your time studying and in your early professional career.

Who (are we, the authors)?

- **Olle Bergman** is a Swedish communication trainer and freelance writer with a background in engineering and science. Additionally, he is the author of a dozen books—mainly non-fiction with a history and language focus—in Swedish.

---

[2] The so-called *Five Ws* are described in Chap. 4.
[3] The industry-associated term *R&D* stands for "Research and Development," that is, the department in a company—big or small—where research, innovation, and engineering take place.
[4] With the expression "science people," we will refer to professionals from the STEMM fields (science, technology, engineering, mathematics, medicine).

- **Sarang Park** is a Korean biochemist, educator, and medical student. She has also been the manager of @IAmSciComm—one of the foremost X (formerly Twitter) forums for science communicators.
- **Joanna Bagniewska** is a Polish-British zoologist and science communicator. She splits her time between academia and the real world, and is an award-winning speaker, a popular science author, and, at times, a stand-up comedian.

(Full bios are found after the table of contents.)

What (does the book contain)?

This book is divided into three parts: *Part I: Forming the Right Mindset*, *Part II: Learning Core Skills*, and *Part III: Addressing Different Tasks, Target Groups and Situations*. Together, these parts aim to achieve the following:

- Introduce communication theory and rhetoric to convey a mindset for effective communication based on time-tested tradition, hands-on experience, and evidence-based research.
- Point out how specific methods, knowledge, and expertise from other domains can be applied to scientific peer-to-peer communication and science outreach.
- Offer a collection of know-how, skills and experience that may help you communicate your data, opinions, and credentials in a way that adapts to the situation and target group at hand. The focus is on spoken and written communication, but we also give introductions to areas like graphic design and video editing.

As you fill the shoes of *The Communicating Scientist*, we want to inspire you to engage more personally and passionately in your professional pursuit of communication.

How (will the authors approach the subject)?

As you soon will note, this handbook observes the communication culture of academia from a skeptical viewpoint. Consequently, we try to apply a more general perspective, bringing knowledge, experience, and trade tricks from professional communicators.

Please note that the title *The Communicating Scientist* emphasizes that the book's scope includes *both* scientific peer-to-peer communication and science outreach. To inspire you, we try to steer away from the conventional style and tone of voice of handbooks of this kind. In addition, we regularly convey tidbits of general knowledge to make the main ideas easier to remember, and the reading experience more fun (occasionally introducing a touch of intrigue and complexity as well).

When (will this book be helpful)?

The general reply is "whenever you take on the persona of *The Communicating Scientist.*" But let's split it into three replies, depending on what situation you're in: the first is "right now," the second is "whenever you need it," and the third—"during the rest of your life."

Preface: A Book That Illuminates, Instructs and Inspires

As a practical handbook, we want this book to be a useful starting point whenever you need to engage in some kind of science-related communication activity. Even if you don't find exactly the protocol you're looking for in a specific situation, reading this book will probably guide you to a solution elsewhere or inspire you to create one of your own. After all, as a scientist, you always need to optimize your protocols.

In addition, our ambition is to help you start thinking differently and acquiring new skills—perhaps even after reading a few pages. In this preface, for example, we have already demonstrated some of the techniques described in the book:

- **Firstly**, we address you directly, dear reader, using a casual writing style. At the same time, we will be careful to serve you a well-structured text, split into easily accessible chunks—such as this bullet list. (Read more about writing techniques in Chap. 8.)
- **Secondly**, with the headers in this preface, we illustrate how asking *Why?*, *Who?*, *What?* … can be used to present something in a brief, clear, and punchy manner (Read more about *The Five Ws* in Chap. 4).
- **Thirdly**, we demonstrate how the act of challenging conventions and using quirkiness may spark curiosity. Hopefully this is one of the reasons why you are still reading the preface of our book (Read about the psychology of communication in Chap. 5).

A note on the use of references

You may have noticed that, unlike many other authors in the field of science, we have chosen a more conversational approach to book writing. This means that—aside from using informal language and links to popular culture—we will not pepper every paragraph with journal references.

Science itself is, of course, based on the scientific method, data collection, and extensive use of references. However, while there is empirical research—based on data, lab results, and questionnaires—that teaches us essential things about how science-related things could be communicated more effectively (for instance, a study from Lund University concluded that informed consent documents for clinical trials needed a better graphic design to be effective[5]), practical communication is still as much of an art as it is science.

For this reason, the number of references is quite limited. Rather than specific scientific results found in a particular article, we convey know-how from our favorite handbooks and articles, tricks of the trade from professionals, and our own experiences. Still, to help the reader continue their knowledge journey, we will offer suggested reading—both books and articles—in most chapters and sections.

---

[5] Dellson, P. (2019). *Patients in Clinical Cancer Trials. Information, Understanding and Decision-Making.* [Doctoral Thesis (compilation), Department of Clinical Sciences, Lund]. Lund University: Faculty of Medicine.

You will also note that we sometimes use Wikipedia definitions. These are often excellent as they are the result of the joint efforts of an anonymous (if not unanimous) team. As we see it, it is time to take Wikipedia seriously.[6]

Happy reading! And please remember: every time you communicate, make sure you know what your core messages are and what effect you are striving for (Fig. 1).

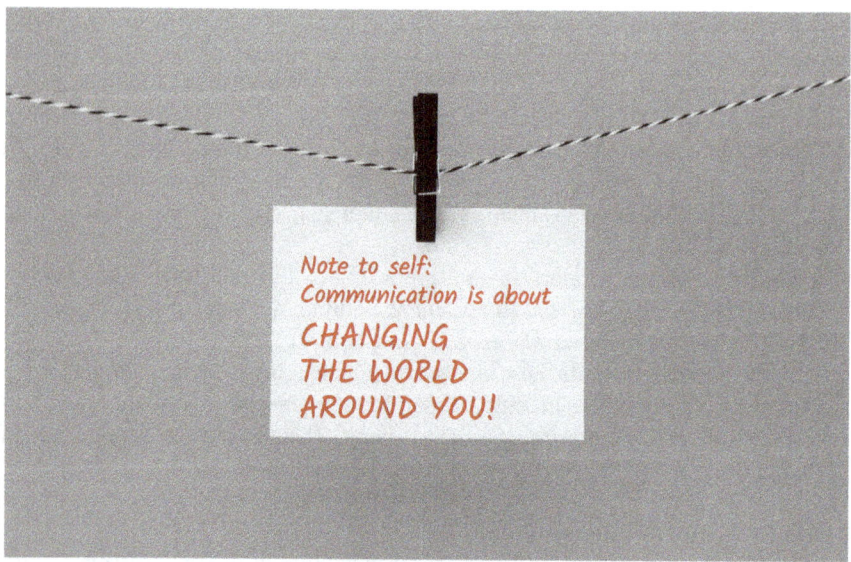

Fig. 1 The purpose of communication is to drive change—in various settings, capacities and contexts. [*Photo by* Kelly Sikkema on Unsplash]

---

[6] There are, of course, innumerable Wikipedia articles of low quality, not least among those which are not written in English. But after a subject has been thoroughly dissected, scrutinized and debated by an international community of nameless editors, the Wikipedia articles often end up superior to any textbook material when it comes to thoroughness, accuracy, balance and declaration of controversy and opposing views.

Acknowledgement and Reference

Thank you to Sam Illingworth, Edinburgh Napier University, for contributing to this chapter.

Dellson, P. (2019). *Patients in Clinical Cancer Trials. Information, Understanding and Decision-Making.* [Doctoral Thesis (compilation), Department of Clinical Sciences, Lund]. Lund University: Faculty of Medicine.

Eskilstuna, Sweden
Shanghai, China
Oxford, UK
2020–2024

Olle Bergman, M.Sc.
Sarang Park
Dr. Joanna Bagniewska

Acknowledgements A number of friends, contacts, and colleagues have read our text, contributing with additions, suggestions and ideas, as well as with corrections, opposing views and constructive criticism. In some places we have been using text snippets—a wording here and a sentence there—directly from their comments. (*Look at that, guys—we've been writing a book together!*)

As such, we would like to thank:

- **Readers and Contributors**: Cecilia Åkesdotter, Tor-Martin Austad, Heather Buschman, Jane Carmichael, Olga Ceran, Helen Cullen, Michael Drout, George Economides, Lydia Flock, Joseph Fridman, Aitor De Andrés González, Ada Grabowska-Zhang, Martina Hysi, Sam Illingworth, Sophia Junker, Matilda Johansson, Johanna Mayer, Natalia Osica, Claire Price, James Ryerson, Peter Ström, Marta Topor, Jessica Vargas, Richard Walters
- **Champions**: the Bergman Family, Chantal Bohn, Håkan Karlsson, Ian Preston
- **Illustrator**: Dorota Paczesniak
- **Cover design**: Anina Takeff

Special thanks to Gonzalo Cordova at Springer Nature who initiated the whole project—thanks for believing in us! "You can't start a fire without a spark" (Bruce Springsteen, "Dancing In the Dark").

This book has been the center of a growing science communication community. If you have contributed to its shape and we have missed your name, we can only apologize—it was not done in malice. We thank you deeply.

Competing Interests The authors have no conflicts of interest to declare that are relevant to the content of this book.

Intro: How to Make Use of This Book

This book is divided into three parts, which will guide you through a three-step process.

Step 1: Forming the Right Mindset

We begin with an exploration of communication theory, drawing on selected theoretical models and the ancient art of rhetoric. This part combines current research with insight from cultural tradition and professional experiences, such as storytelling and sales. We also address human psychology and dwell on subjects like cognitive dissonance and biases. This foundation allows us to draw on best practices from areas such as corporate communications, publishing, journalism and opinion-making throughout the book.

Step 2: Acquiring Core Skills

The middle part of the book presents core skills and techniques for addressing communication tasks essential for science and tech people. Such techniques include the rhetorical elements ethos/pathos/logos, journalistic writing techniques, Hollywood-style storytelling, and more.

Step 3: Approaching Different Tasks and Situations

The final part of this book is a task-related catalog of communication activities and situations, covering various modes, channels and tools. This includes everything from oral presentations, writing grants, scientific poster production and slide design, to media interactions, networking activities and building and online presence (Fig. 2).

Fig. 2 Metaphorically, this communication handbook can be compared to a glorified cookbook, where Part I is about nutrition and physiology, Part II contains recipes for the weekdays, and Part III offers recipes for special occasions. [*Photo by* Micheile Henderson on Unsplash.]

A Communication Crib Sheet for Science People

The Central Themes and Advice from This Book Summarized for Your Convenience

> **What you will learn from this chapter**
>
> We begin this book with what we think are the most important themes for you to grasp, understand and remember. Our list of essentials conveys some basic methods, principles and ways of thinking that are universal enough to help you plan and execute a wide range of communication activities. Treat this chapter as the essence of our communication expertise and advice.

As you read *The Communicating Scientist*, you will notice that we keep returning to a few key themes. We do so to hammer in a set of powerful principles that can be used universally in a range of situations. After all, communication has the same nature as, say, molecular biology, wave physics, or linear algebra: to master the discipline, you have to understand and skillfully apply a set of core concepts. In the crib sheet below, we summarize the most important of these principles into twelve rules.

(We define *communication activities* as any action taken to convey information, attitudes, or emotions. This includes everything from personal conversations and simple emails to scientific presentations, journal articles, research posters or YouTube videos.)

Rule #1: Define your core messages and let them lead you.

What does that mean?
The planning, preparation, and execution of any communication activity should revolve around the main messages you want to convey.

How is it applied?
Structure all communication around your key messages, in a way that makes them clear to the receiver. Start with the question, "WHAT am I setting out to convey?"

The rest of the content serves to provide context, explain how things are connected, and offer relevant details.

Your main messages can be delivered in different ways, depending on the channel, format, and composition:

- at the start ("frontloading"—extensively used in news media),
- at the end (often seen in stories, for instance in the form of a moral), or
- throughout the course of the communication activity.

Remember: if you don't know what you are trying to say, don't expect the audience to hear it. Or read it. Or see it.

Rule #2: Ensure every communication activity has a clear purpose.

What does that mean?
If a communication activity lacks a well-defined purpose, it is questionable whether it has any value at all. Skilled communicators always aspire to introduce change with their actions, or influence their audience in some way.[1]

How is it applied?
When planning your communication activity, identify clear targets; the question "WHY am I doing this?" will help you recognise them. Your goals may include conveying information, creating understanding, winning someone over to a cause, or selling something (a product or an idea). Follow the motto: "Let's change the world a little!"

Rule #3: Tailor your activities to the context, the medium, the target group, and yourself.

What does that mean?
Every communication activity operates within a framework, and is therefore bound by a set of constraints. Consider, for instance:

- Context—the background, setting and goal.
- Channel—the medium used.
- Receiver—the target audience.
- Sender—your strengths and weaknesses (but also how you or your team are perceived by the target group).

---

[1] An engineer could express it like this: "State A of the audience before the communication activity should change into State B after the activity." Or to note it in an even nerdier way: $Audience_{before}$ => $Audience_{after}$.

How is it applied?
When preparing a communication activity, analyze its conditions and limitations. Do so by asking the following question about each of the above factors:

- What is the context/channel/receiver, etc., and what is special about it in this particular case?
- How does it affect my situation?
- With this knowledge, how do I tailor the communication activity to reach my goals as effectively as possible?

Rule #4: Adapt to the needs and expectations of each audience.

What does that mean?
The audiences you meet as a scientist—university professors, business investors, journalists, and so on—have very different ways of defining successful communication outcomes. The principle of "one size fits all" does not apply; instead, you must adapt your mindset and your communication activities to each context and target group.

How is it applied?
The key to success lies, firstly, in thorough preparation, secondly—in your perceptiveness. Do your homework—read, ask questions, discuss—to understand who your target group is and what it wants from you; then design your communication activity based on what you have found out. If, for instance, you were to give a lecture about vaccine research, you'd emphasize different points when addressing pediatricians, compared to a vaccine-hesitant audience. While you interact, set aside prejudices and preconceived notions, listen more than you talk, and aim to serve your audience in the most sincere way.

Rule #5: Treat communication as the lifeblood of all your professional and social activities.

What does that mean?
Human activities, cooperation and socializing are based on mutual communication—all the time, everywhere. The evolution of *Homo sapiens* enabled rich, complex and continuous communication through many channels; this ability laid the groundwork for all manifestations of human progress. Failures—be it social, organizational, or technical—are often a consequence of suboptimal communication.

How is it applied?

By becoming aware of how and when information, opinions, or attitudes are conveyed in different ways between individuals and groups, you increase your chances to

- work more effectively,
- succeed professionally,
- increase your influence,
- maintain good relationships.

On top of that, the practical skills you learn and develop (through this book and beyond) will allow you to communicate more efficiently and in new ways.

Rule #6: Understand and master the ways in which science can be communicated.

What does that mean?

Science may be communicated for a number of purposes, to a wide array of target groups, and using different types of channels, genres, techniques and strategies. However, communication activities related to science normally belong to at least one of these five categories:

- Peer-to-peer communication,
- Public outreach,
- Strategic or commercial communication,
- Science education,
- General professional and social communication.

How is it applied?

Each of the five categories represents its own communication culture, based on different goals, messages and target groups; therefore, the same set of scientific information may be communicated in very different ways. Still, many communication skills are generally transferable, and work across these cultures.

Rule #7: Use knowledge of rhetoric and modern psychology to increase the effectiveness and impact of your communication activities.

What does that mean?
Communication is generally more effective when based on an understanding of "human nature"—that is, the psychology of motivation, interpersonal behavior, attitudes, self-image, and so on. This area can be studied from two angles: ancient rhetoric and modern psychology.

How is it applied?
Rhetoric, "the art of persuasion," offers a number of well-proven techniques. Here are three examples:

- The Aristotelian triad of *ethos* (trustworthiness), *pathos* (emotion), and *logos* (facts and logic) guides speakers who intend to persuade.
- The Five Ws (What?, When?, Where?, Who?, Why?) identify most relevant information.
- The universal "rule of three" (our brains expecting a collection of three elements) helps to structure texts and talks.

By studying how minds work, modern psychology reveals key knowledge useful for communicators. For instance:

- Research on **cognitive bias** gives important clues on how humans can misinterpret and distort information they receive; in some instances, they may be reluctant to accept certain facts.
- The principle of **salience**—from perception psychology—explains why an audience is more motivated to listen to a communicator who presents something that is out of the ordinary in some way.

Rule #8: If the context permits, address the audience directly, using plain language and a conversational style.

What does that mean?
Abstract language increases the distance between sender and receiver. The most effective way to reach an audience is to be inspired by the informal dialogue between a skilled teacher and students—an interaction based on questions, explanations and mutual curiosity.

How is it applied?
If the format allows, avoid passive voice in writing. Instead, address the reader directly and make it clear who the sender of the message is (like we are doing

here, right now). Use a conversational style to break complex subjects into understandable chunks. Strive to use the same principle during presentations, engaging the audience in a person-to-person dialogue.

Rule #9: Design presentations to convey a fixed set of key messages.

What does that mean?
Presentations often take place where there is fierce competition for attention and a massive flow of information. Thus, it's essential to ensure that your main points are clearly communicated and not lost in a multitude of less important details.

How is it applied?
Start the preparation work by defining the goal and key messages of your presentation. Let every line of the script and every slide serve these main messages. Reduce detail; those interested will seek more information afterwards—during the Q&A, through personal contact, or in published literature. Ask yourself: "If my audience only remembers three things from my presentation, what should they be?"

Rule #10: Design scientific posters to serve as an advertisement and a networking tool.

What does that mean?
A scientific conference poster is not a journal article splayed out on a wall. Instead, its purpose is to prompt curiosity, questions and dialogue, ultimately aiming to create relations with peers and other professionals.

How is it applied?
Design your poster to tell a visual story, focusing on conclusions, main results, and key methods. Include comprehensive information on how to get in touch with you—in person, or via social media and email.

Rule #11: Good writing is writing that does its job.

What does that mean?
The most effective way of writing a text can vary considerably depending on the purpose, channel and audience. A skilled writer knows how to adapt the genre, composition, style, and tone of voice accordingly.

How it is applied?
By constantly practicing writing for different purposes and channels, any writer can improve their skills to produce text that is effective in its context and "does its job." Here are some practical guidelines useful for any type of text:

- **Well-crafted paragraphs**, which begin with a topic sentence, are key components of quality writing.
- **Signposting**—using connecting devices like "therefore," "however," "in conclusion" to guide the reader—clarifies the structure of the text, and links sentences, paragraphs, and ideas.
- **Chunking**—splitting the text into smaller units to avoid walls of words—makes reading faster and understanding easier. It also motivates the reader to explore the writing in more detail.

Rule #12: Strategize—prepare, execute, coordinate and develop your communication activities according to a plan.

What does that mean?
In the domain of science, communication activities are often planned and performed on an *ad hoc*[2] basis, not seen as parts of a larger scheme. By introducing strategic thinking about communication, scientists can find new ways to reach their goals faster and with less effort.

How is it applied?
Set up an integrated communication plan to define short- and long-term goals, spot strengths and weaknesses, coordinate different activities, and develop a road map. It is often wise to define the limitations (such as time, resources and skills) before exploring the possibilities.

---

[2] *Ad hoc* is a Latin term that is useful to learn. It translates to "for this" in English. It usually refers to a solution that is adapted to a specific purpose, problem or task, rather than a general solution that can be applied to multiple similar situations.

In Memoriam

Olle Bergman, 1964–2023

In December 2023, our beloved Olle, husband and father and grandfather, went on his last orienteering run. His heart stopped as he crossed a dirt road, aiming to reach a small creek beside a large field. He was rushed to hospital, but his life could not be saved. As a family, we find some comfort in the fact that he died doing what he loved best: running free, at one with nature, surrounded by beautiful scenery, and with cool air filling his strong, healthy lungs.

Olle was passionate about many things. The book you're about to explore was a dream of his, aiming to help people in the two areas that excited him most—communication and science. Olle spent so much of his time helping others, especially young science and tech professionals. He thrived in the company of others, and generously shared his knowledge and experience with anyone who was interested.

We are immensely proud of this book and the efforts that have been put into it. Above all, we are extremely grateful that Olle's life-long science and communication work continues to live on. Thank you to all who have inspired, challenged, applauded, and cared for Olle. He truly was a people person, and so many have helped to fill his life with joy, music, adventure, history, literature, nature, cultural and creative work, and inspiration.

Lotten, Erik, Ida and Julián, Oskar, Moa, Sigge, Freja and Matilda

Contents

Part I Forming the Right Mindset

1 Conveying the Observations, Insights and Wonders of Science 3
  1.1 A Significant Part of Your Weekly Work is Spent on Communication Activities 3
  1.2 Communication is an Integral Part of Science 4
  1.3 Today, Communicating Science Falls into Five Main Categories 8
  Reference 12

2 Communication—The Lifeblood of *Homo Sapiens* 13
  2.1 Communication—A Universal Concept 13
  2.2 Modes of Communication 15
    2.2.1 Suggesting a Seven Stage Model 15
    2.2.2 Learning from the Seven-Stage Model 19
  2.3 Three Observations Regarding Communication 20
  References 24

3 Several Handy Theoretical Models for Practical Communication 25
  3.1 Introducing a Bit of Theory 25
    3.1.1 Let's Define the Concepts We Use 26
    3.1.2 Let's also Agree on the Terminology 26
  3.2 A Handful of Useful Communication Models 28
    3.2.1 The Shannon-Weaver Model—The Boss of All Communication Models 29
    3.2.2 Auxiliary Models 32
  References 35

4 Understanding the Universal Usefulness of Rhetoric 37
  4.1 What is Rhetoric All About? 37
  4.2 Rhetoric is Everywhere 38
  4.3 Some Glimpses into the History of Rhetoric 40

xxxiii

	4.4	Putting Rhetoric to Use	43
		4.4.1 The Aristotelian Triad: Ethos, Pathos Logos	43
		4.4.2 The Canons of Rhetoric	45
		4.4.3 The Five Ws	46
		4.4.4 Other Important Principles of Rhetoric	47
	References		49
5	**Some Notes on the Psychology of Human Communication**		51
	5.1	How Does Psychology Link to Communication?	51
	5.2	Reflecting on Personality	53
		5.2.1 The Importance of Trust	54
	5.3	What Gets in the Way of Communication	55
		5.3.1 Spontaneous Trait Inference	55
		5.3.2 Confirmation Bias	56
		5.3.3 Cognitive Dissonance	58
		5.3.4 Dunning-Kruger Effect	59
	5.4	Memory, Attention, and Retention	60
		5.4.1 Directed Attention—Imperative in All Kinds of Communication	62
	References		66

Part II Learning Core Skills

6	**Finding the Right Starting Point for Any Communication Activity**		71
	6.1	Setting Up a Plan	71
	6.2	Five Universal Guidelines and How to Use Them	72
	6.3	Putting the Rhetorical Elements Ethos, Pathos, Logos to Use	78
	6.4	Telling a Good Story	82
	References		88
7	**Planning, Preparing, and Performing Persuasive Presentations**		89
	7.1	A Recipe for a Successful Speech	89
		7.1.1 The Power of the Spoken Word	90
		7.1.2 Becoming a Great Speaker	91
	7.2	General Preparations for Any Talk	94
	7.3	Presentation Delivery	99
		7.3.1 Using Your Body to Communicate	100
		7.3.2 Using the Space Around You to Communicate	105
	7.4	Making Everyone Feel Safe	108
	7.5	Preparing for a Q&A	109
	7.6	Managing Your Nerves	110
		7.6.1 Practical Tips for Reducing Stress	113
		7.6.2 Handling Disturbances	114
		7.6.3 Some Useful Precautions	116
	References		116

8 Producing Quality Writing for a Range of Purposes 119
- 8.1 A Note to the Human Writer, Considering the Robot Writer ... 119
- 8.2 Skilled Writers Have an Amazing Tool of Persuasion at Their Hands 120
  - 8.2.1 Defining Good Writing as Effective Writing 121
- 8.3 The Craft of Writing 122
  - 8.3.1 The Essence of Good Writing 123
  - 8.3.2 Making Your Writing Do Its Job 125
- 8.4 Approaching a Text Assignment 128
  - 8.4.1 Starting with the Summary 128
  - 8.4.2 Brainstorming + Outline 129
  - 8.4.3 Orientation, Information and Action 130
  - 8.4.4 Draw an Article 130
  - 8.4.5 Managing Writer's Block 131
- 8.5 Some Tricks of the Trade 133
  - 8.5.1 Chunking—A Cornerstone of Modern Writing 134
  - 8.5.2 "The Inverted Pyramid" 135
- 8.6 Generative AI in Science Communication 138
  - 8.6.1 Addressing Concerns Over the Use of GenAI 138
  - 8.6.2 Potential Benefits of Using GenAI 140
- References 141

9 Designing Effective Visuals 143
- 9.1 Visual Communication is Essential Communication 143
- 9.2 Scientists Who Think in Design 144
- 9.3 The Fundamentals of Design in Science 146
  - 9.3.1 The Appeal, Comprehension, Retention Model 146
  - 9.3.2 The CARP Model 147
  - 9.3.3 Zen Faulkes' Similarity and Contrast Model 149
- 9.4 Guidelines for Reluctant Visual Designers 150
- 9.5 Visual Hierarchy Helps Us Organize Information 155
- 9.6 Making Typography Work for You 157
- 9.7 Choosing and Managing Colors 161
- References 168

Part III Addressing Different Tasks, Target Groups and Situations

10 Instructions for Speaking in Different Settings 173
- 10.1 Introduction 173
- 10.2 The One-Minute Elevator Pitch 173
- 10.3 Marketing an Idea, Project or Product 178
- 10.4 Participating in a Panel 182

	10.5	Giving Longer Talks and Workshops (1–6 h)	186
	10.6	Planning and Executing a Teaching Session	189
	10.7	Turning a Live Lecture into an Online Lesson	193
	10.8	Giving a Media Interview	197
	10.9	Giving an Official Speech or Informal Toast	202
	References		204
11	**Instructions for Different Writing Tasks**		**205**
	11.1	Introduction	205
	11.2	Articles for Peer-Reviewed, Scientific Journals	205
	11.3	Writing Effective (One-on-One) Emails	214
	11.4	Grant Proposals	219
	11.5	Press Releases	223
	11.6	Writing for Mainstream Media	227
	11.7	Writing an Interview for a Magazine, Newsletter or Blog	232
	References		235
12	**Some Tips for Visual Communication**		**237**
	12.1	Intro	237
	12.2	A Slide Deck for a Short Scientific Presentation	237
	12.3	Designing an Effective Scientific Poster	242
	12.4	Producing Video Material, Such as an Abstract	249
	References		255
13	**Two-Way Communication in Various Contexts**		**257**
	13.1	Introduction	257
	13.2	Networking	258
	13.3	Presenting in an Industry Setting	262
	13.4	Public Engagement	265
	13.5	Building a Relationship with Mainstream Media	272
	13.6	Addressing Skeptical Audiences	276
	References		279
14	**Setting up and Executing a Plan for Your Online Presence**		**281**
	14.1	Introduction	281
	14.2	Using Social Media Effectively	283
	14.3	Creating a Personal Website	288
	References		291

Epilogue: A Communication Manifesto for Science ... 293

Further Reading ... 295

Authors and Illustrator

About the Authors

Olle Bergman is a Swedish communication trainer, freelance writer, and author with a background in engineering and science.

He received his degree in Chemical Engineering from the Faculty of Engineering, Lund University, in 1989. After spending three years in a wet lab context, working as a research assistant in neurochemistry and molecular biology, he entered a communications career. During some intense years in different professional positions, he learned the nuts and bolts of corporate and academic communication, as well as publishing and advertising, in a most practical way.

After becoming a freelancer in the late nineties, Olle Bergman has mainly served customers in science, medicine, and tech. Typically, he helps his clients with copywriting and strategic planning, as well as coaching and lecturing on practical communication. Among his customers through the years are organizations like automotive manufacturer Scania, the Swedish Cancer Society as well as a number of universities, e.g. Lund University, KTH Royal Institute of Technology, and Karolinska Institutet.

In addition, Olle Bergman is the author of a dozen book titles—mainly history—and language-oriented non-fiction—in Swedish. As a member of the scholarship board of *The Swedish Authors' Fund* he has been dwelling on the question "How do you recognize a well-written text, and how should we define text quality?"

Olle passed away in December 2023.

Sarang Park received her degree in Biochemistry and Cell Biology from Jacobs University Bremen in Bremen, Germany, in 2019. She played a central role in developing academic life for students, such as spearheading the Biochemistry and Cell Biology Society as its President and resuscitating the Jacobs University Bremen Undergraduate Newspaper as the Co-Editor-in-Chief. She has also led numerous projects such as jacobsHack! and TEDxJacobsUniversity. Throughout, she has accumulated eight years of teaching experience aiding numerous students and cultivating a love of learning as a freelancer and a consultant. Simultaneously, she acts as the social media manager of the largest science communication X (formerly Twitter) community @IAmSciComm with 38K followers.

Currently, she is attending medical school at Shanghai Medical College, Fudan University, in Shanghai, China.

Dr. Joanna Bagniewska completed her undergraduate degree at Jacobs University Bremen and Rice University in Houston, and obtained her M.Sc. and doctorate from Oxford University's Wildlife Conservation Research Unit. After a stint at a start-up company, where she trained bees to detect illegal substances, Joanna went on to lecture at Nottingham Trent University, the University of Reading and Brunel University of London. Her academic interests include conservation biology behavioral ecology and the intersection of technology and zoology. Joanna has worked on a number of species ranging from wombats and wallabies to mole-rats and jackals.

She is an accomplished science communicator, having won British Council's FameLab Poland and the Wellcome-funded *I'm a Scientist, get me out of here!*, given a TEDx talk, and performed at science stand-up comedy events. She spent six years working as a Communications and Public Engagement Officer at Oxford University's Department of Paediatrics, and has been a science communication coach for the British Council. As a freelancer, she regularly writes popular science articles for various media in Polish (including Focus, Gazeta Wyborcza and Tygodnik Powszechny) and English (including TLS and BBC Wildlife), and has collaborated with the Discovery Channel on the *How Do They Do It?* series. Her first popular science book, *The Modern Bestiary*, was published in 2022.

Joanna works as Co-Director of the Postgraduate Certificate in Ecological Survey Techniques at Oxford University's Department for Continuing Education.

About the Illustrator

Dr. Dorota Paczesniak is a freelance scientific illustrator, educator, and an evolutionary biologist.

She studied biology and geography at the Jagiellonian University in Kraków, Poland. She completed her Ph.D. at ETH Zürich, Switzerland, focusing her research on asexual reproduction in wild snail populations. She then continued working as a researcher at the Leibniz Institute of Plant Genetics and Crop Plant Research (IPK) in Germany and the University of Saskatchewan in Canada to study evolutionary questions related to reproductive systems in plants. She has worked as a lecturer at ETH Zürich, Switzerland, and the University of Oulu, Finland. She is an author of popular science articles and educational materials about evolution, genetics and plant breeding.

As an illustrator, Dorota has designed a series of infographics about forest tree breeding for a public information campaign of the Natural Resources Institute Finland (Luke) and the University of Helsinki, illustrated a popular science book, and worked with many academic researchers to develop figures for scientific publications.

Part I
Forming the Right Mindset

Part I of this handbook will help you shape a constructive mindset towards communication—a way of thinking that is practical, results-oriented, and people-focused. It merges traditional, pre-IT wisdom (from areas such as rhetoric, media, and marketing) with contemporary, internet-savvy proficiency (from areas like social media and popular culture). Additionally, we throw in some modern cognitive psychology.

Grasping the more philosophical themes of *Part I* will help you implement the hands-on instructions and advice of *Part II* and *Part III* in a more effective way.

Conveying the Observations, Insights and Wonders of Science

Why Science Can be Defined as a Communication-Based Operation

> **What You Will Learn from This Chapter**
> Communication has a central place in all human affairs, and science is no exception. However peculiar it may seem, the chief part of an average work week for a scientist is actually spent on communication-related activities.
> While reading this chapter, you will:
>
> - explore how most activities in science are inherently dependent or centered around interhuman communication.
> - dwell on the target groups and audiences a scientist interacts with today (hint: there are more of them than you may think).
> - examine the five categories of communication-related activities in science: scientific peer-to-peer communication, public outreach, strategic communication, scientific education, and general professional and social communication.

1.1 A Significant Part of Your Weekly Work is Spent on Communication Activities

How would you respond if someone asked how much time you—as a scientist—spend on different communication activities during an average work week?

Based on our anecdotal experience, few scientists can offer a straight-forward answer to this question. Firstly, the concept itself causes some bewilderment: what exactly counts as "communication"? Secondly, while scrutinizing the question, one realizes that "communication" is not only hard to define, but also entangled with many other activities.

Here's what we think the answer is: "A lot more than most scientists realize!" In fact, there is an undercurrent of communication flowing through all science-related endeavors—just take a look at Fig. 1.1.

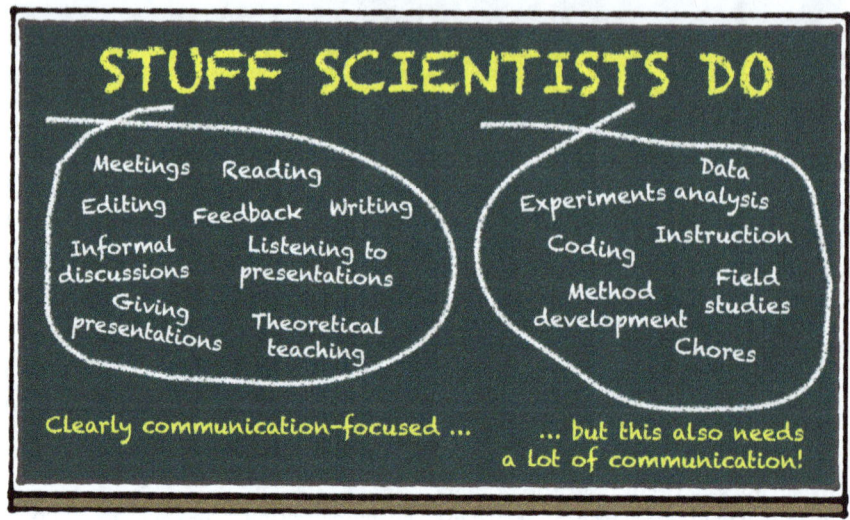

Fig. 1.1 All activities in science are inherently dependent or centered on interhuman communication; the exceptions would be instances of lonely lab-or fieldwork. It is probably easier to point out the few activities that do not include communication in some form

> **Reflection 1.1**
> Write down a typical weekly schedule which includes the categories in the illustration above. Make sure to include informal but work-related conversations and your reading work. What is your take on our estimate? Is the majority of your average working week actually spent on communication-related activities?

1.2 Communication is an Integral Part of Science

If science weren't communicated, it wouldn't exist as we know it today. Without communication, some core concepts of the scientific process—such as reproducibility and peer review, as well as the recording and sharing of data—would be meaningless.[1] Communication and research are two concepts that have existed side by side throughout history.

> **Some Events in the History of Scientific Communication**
> Scientific communication is as old as organized science; below we list some prominent and seminal examples.
>
> **Fourth century** Aristotle compiles ideas, analysis and knowledge from the Golden Age of Greece; his writings serve as foundation stones not only for science as we apply it today (e.g., *Physica*), but also for modern communication (e.g., *Ars Rhetorica*).[2]
>
> **1584** The cosmological theorist Giordano Bruno publishes some of his most important work in response to colleagues like Copernicus and Tycho Brahe. Eventually, in 1600, he is burned alive for maintaining his ideas.
>
> **1660** A learned society is formed in London which becomes *The Royal Society of London for Improving Natural Knowledge*. In 1666, they start publishing *Philosophical Transactions*—the world's first science journal.
>
> **1870–71** The Franco-Prussian War drives a wedge into the communication—and potential cooperation—between Louis Pasteur (French) and Robert Koch (German). Both of them are on the verge of a scientific breakthrough about the link between disease and microorganisms. Their frosty relationship almost undoubtedly delayed the practical application of their joint research.
>
> **1962** Rachel Carson's popular science book *Silent Spring* has an enormous impact; it is considered one of the most important titles for the foundation of today's environmental movement.
>
> **1991** Beginnings of arXiv, the first repository of preprint articles. Started in the pre-WWW era by Paul Ginsparg, arXiv became a pioneering player in the open access movement. Its creation and success prompted the development of resources such as PubMedCentral, publishers like PLoS, and daughter preprint servers such as bioRxiv and medRxiv.
>
> **2016** Ed Hawkins produces "climate stripes"[3]—a minimalist yet very powerful visualization of long-term temperature trends, ranging from blue (cooler years) to red (warmer years). This simple but evocative visual has been adapted for different geographical locations, and reproduced on merchandise, infrastructure and fashion items, becoming an instantly recognisable global phenomenon (Fig. 1.2).

---

[1] A lonely basement-bound researcher could of course make some progress during his or her lifetime. But then all their results and conclusions would be left to oblivion.
[2] Some of his less accurate ideas—like the false belief that hedgehogs carry apples on their spines to store for later—have been dragging on in popular culture for centuries. Now THAT'S effective science (well, "science") communication!
[3] Hawkins (2023).

Fig. 1.2 Warming stripes. The progression from blue (cooler) to red (warmer) stripes portrays annual increases of global average temperature since 1850 (left side of graphic) until 2023 (right side). *Source* https://showyourstripes.info/

Throughout history, the communication aspect of science has relied upon a few key activity types:

- writing (and reading)
- speaking (and listening)
- creating and showing visuals (and examining them)
- demonstrating experiments and specimens (and attending such demos).

Applying the bullets above to our times, the typical and expected communication skills of a scientist are thus as follows[4]:

- writing journal articles, reports or applications
- keeping up with daily correspondence—mostly emails to recipients near and far
- giving presentations
- engaging actively and purposefully in conversations and discussions

---

[4] In addition, there is of course a receiver's perspective to the list: scientists also have to achieve a high degree of perception and understanding. This is not only about the texts they read and the presentations they follow—it is also about the social skills necessary to have a conversation or discussion that is meaningful and useful for all parties involved.

1.2 Communication is an Integral Part of Science

- designing scientific figures and illustrations, presentation slides, and research posters
- demonstrating experiments and specimens (though this is more common in science teaching than in peer-to-peer communication).

Now, please have a look at the cover of this book. Note that it doesn't say "scientific communication" or "science communication." Instead, by using the title *The Communicating Scientist*, we have rooted it in the perspective established above: that most scientific activities heavily depend on interpersonal communication. As a result, we offer a very broad approach to the communication activities a scientist engages in—some daily, some monthly, some yearly, and some only on special occasions. As you will see, our practical instructions include a wide range of purposes, genres, channels and target groups (Fig. 1.3).

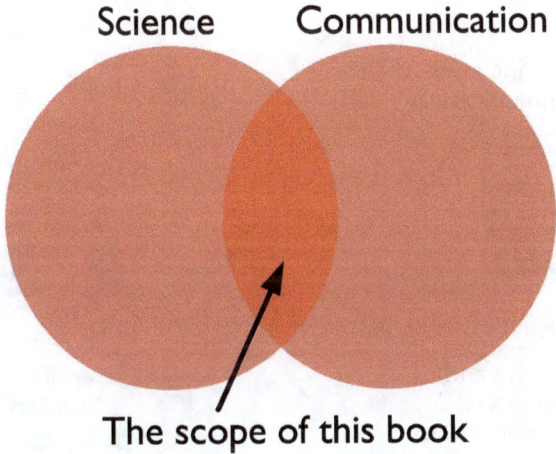

Fig. 1.3 There is so much more than journal articles, presentations, and scientific posters to explore in the intersection between science and communication

---

Reflection I.II
Write down ten things that can be included in the intersection between the concepts of Science and Communication. We'll start with some examples:

- An article in Nature.
- The regular Monday meeting with your research group.
- The online competition "Dance your Ph.D.".
- You explaining what you do in your lab to your grandma.
- An op-ed in Die Zeit on wind energy.
- The EU Parliament COVID inquiry.

1.3 Today, Communicating Science Falls into Five Main Categories

Over the past few decades, the academic landscape has become far more complex than it was for our scientific forebears. Gone are the days when academics could bring a cushion to their ivory towers and get comfortable up there. Today, we inhabit a supremely connected planet, facing global challenges alongside professionals and laypeople of all backgrounds. As a result, scientists not only *can* but also *need* to communicate facts, ideas and opinions to a broadening array of target groups, and in a range of varied contexts. Still, we think that all communication activities of a scientist can fit into one or more of five categories (summarized in Fig. 1.4).

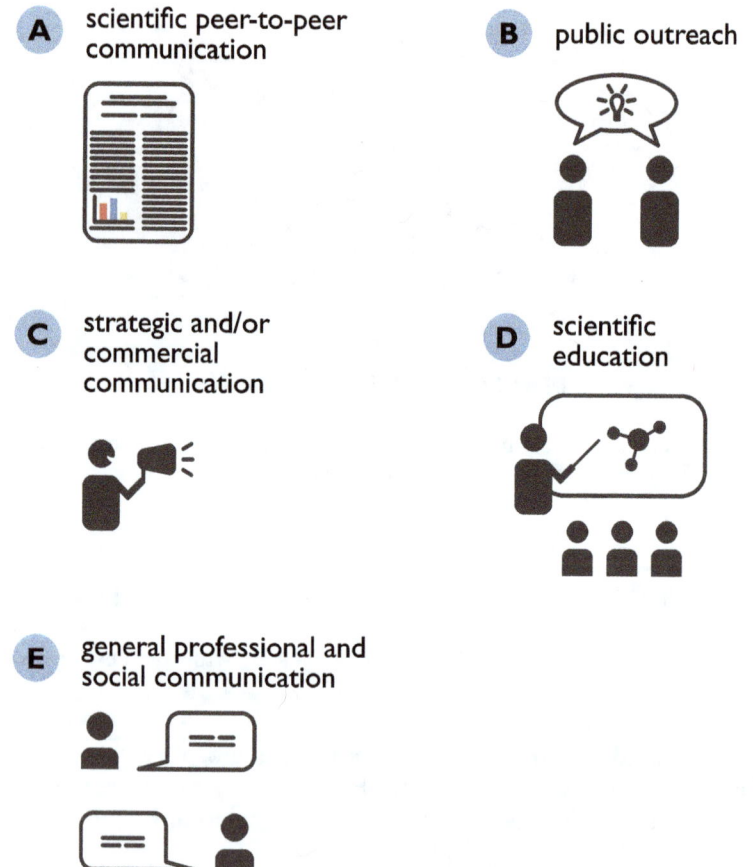

Fig. 1.4 A summary of the five categories of scientific communication

1.3 Today, Communicating Science Falls into Five Main Categories

So, what are they, and which audiences do they address?

Category A: Scientific Peer-to-Peer Communication
This type of communication serves the progress of science and mainly comprises experts interacting with each other, for example via journal articles, oral presentations, scientific posters, or grant applications.

Audience: the scientific community. For most scientists, their peers will—for obvious reasons—always be the number one focus. This group can be divided into three sub-groups:

- "The usual suspects," that is, scientists you know by name, who work in the same field as your research group. Some may be partners, others competitors.
- Scientists in your broader field, who have the specialized knowledge to understand and be interested in your research.
- Scientists who are working in other fields, but for one reason or another are interested in what you do.

Category B: Public Outreach
Also referred to as *science dissemination, science communication,* or *SciComm,* this category encompasses scientists communicating with the outside world. In the lucky instances where a two-way process can be initiated, *public engagement* is a more appropriate term. Science outreach promotes public awareness, understanding, and appreciation of science. While its motivation is often altruistic and conveys the wonders of nature, in many countries disseminating research is also required by grant agencies, who mandate scientists to make certain efforts to engage with the wider public.

Audience: the public. Outreach is typically performed by active scientists or dedicated science communicators, addressing audiences of all kinds—from school children to parliamentarians. If your science matters to you, we assume that you'll want everyone else to share this opinion. To be convinced, they need not only your facts, but also your trustworthiness and engagement.

Audience: the media. Having your research disseminated is generally a good thing, and the more you engage with the media, the greater your potential to control how your information is presented. The first step is to be responsive when the media reach out to you. The second step is to take the initiative to reach out to them. As your skills develop, you will eventually be able to (partly) influence where and when your stuff ends up in their news feed.

Some universities—though this varies widely between countries—have modern communication departments eager to assist their scientists, scholars and students with media presence. After all, a wider reach often translates to a greater research impact. Your university's media team will love to build a strong relationship with you and collaborate on producing media-friendly content.

Category C: Strategic and Commercial Communication
Although rarely discussed in science communication handbooks, activities in this category take place all the time, as scientists recognize the need to communicate for targeted outcomes. Precise purposes vary widely and depend on the nature of the professional organization initiating the communication; they can be commercial, related to public image, political, journalistic, or otherwise. Examples include:

- A startup company communicating the science behind their product to attract investors.
- A major med-tech company keeping a tight rein on the flow of information about their products for different target groups.
- A university promoting its most successful scientists to attract the attention of, for example, media, funders, senior researchers and young talent.
- An environmental charity explaining the mechanisms behind climate change to mobilize public support and influence political decisions.
- A diagnosis-centered medical society advocating for practitioner-based policies to improve the healthcare system.

Audience: Research Funders. While funds typically enter your department via grants—either from government programs or non-governmental organizations (NGOs) and foundations—business investors might also be interested in financing your work. To attract their support, it's crucial to consistently produce good results and build a solid track record; this demonstrates that their investments are worthwhile. Take every opportunity to raise the profile of your research unit.

Audience: Politicians, Officials, Public Agencies and NGOs. As humankind faces bigger and more complex problems (from biodiversity crisis to aging societies), science experts need to inform, influence and empower the decision-makers of their nations and the planet. Additionally, scientists may—for their own sake—choose to engage and influence decision-makers to create better conditions for academia and research.

Category D: Scientific Education (Scientists and STEMM Teachers Teaching Their Students)
Teaching—from preschool to Ph.D. studies—is an essential part of the complex communication landscape. However, it is not strictly within the scope of this book; hence we will largely bypass the educational domain (with a few exceptions).

Audience: Students near and far. With the rapid build-up of knowledge, there is increasing pressure from students to supplement their learning with fresh news from the research front. This is a good thing, as piping-hot case studies from science in progress are a great way to motivate students and demonstrate how research works. Also, when recruiting to your team, remember that the top students and staff of tomorrow are out there, reading science news, following social media, and watching video presentations—honing their theoretical and practical skills as they do so. Some of them might already be at your institution.

Category E: General Professional and Social Communication
This final category collects the communication "everyone" engages in to function as a professional in modern society, which is not encompassed by the categories above. This includes writing emails, composing job applications and cover letters, participating in interviews, and so on. In Part III, you will see that there is often a "science twist" to parts of these tasks.

Audience: Pretty Much Everyone and Anyone

> **Some Pet Peeves of a Communications Trainer—and How to Get Rid of Them**
> *A personal reflection by Olle Bergman*
>
> For 30 years, I have been working in the professional area of communication and training, crossing the borders between different professional domains—corporate communication, advertising, PR, media, and so forth. Although I have finally (re)found my preferred tribe among science people, I am still bothered by some of the shortcomings of the communication culture of STEMM faculties. To sum up, I think these are the main problems:
>
> - **Lack of strategic approach**
>   Communication is seen as something that's "just there," not as something that you can develop and streamline.
>
> - **A conservative mindset with regards to using different channels**
>   Most scientists and groups center on what the culture dictates: talks, journal articles and posters. The use of other channels still varies widely (but things are getting better—just look at you reading this book!).
>
> - **A widespread DIY culture and few interactions with professional communicators**
>   When it comes to know-how, prestige and financial turnover, an ambitious research group is equivalent to a mid-size company. Still, most of them are caught in a "we-can-handle-this-ourselves" mindset regarding communication—reluctant to engage communication consultants or sub-suppliers of specialized communication services.

- **Lack of systematic and formalized training**
  Sure, there are presentation techniques workshops and poster design clinics, but rarely proper specialization and ambition for professional-level skills. Many communication skills are taught and learned "on the fly" during the daily business at the department; this is both good and bad.

Most of these flaws can be addressed through a very simple action program:

Step 1: Take an afternoon off with the research group and run a communication workshop where you address the following:

- Long term goals
- Short term goals
- Main messages
- Channels
- Responsibilities
- Need for external support and services
- Training needs

Step 2: Write a communication plan (see Sect. 6.1).

Step 3: Start working with external experts, for instance in the following fields:

- Design and illustration
- Outreach and strategy
- Training

Step 4: Initiate a training program which consists of two parts: *Basic training* for everybody, and *Individual training*, based on specific needs and group roles.

Reference

Hawkins, E. (2023). *Show your stripes*. Institute for Environmental Analytics. Retrieved July 1, 2024, from https://showyourstripes.info/

Communication—The Lifeblood of *Homo Sapiens*

A Historical and Philosophical Contemplation of the Importance and Power of Communication

> **What you will learn from this chapter**
> There's no doubt that interhuman communication has shaped the destiny of our species. Thus, to become more effective communicators, we should build on a foundation of historical and philosophical insight. As human beings, we are wired to both seek an understanding of the context we are part of, and to feel understood ourselves. Consequently, it is impossible for us *not* to communicate.
>
> Here, you will:
>
> - ascertain the nuances of the word *communication*;
> - review the history of communication through a seven-stage model; and
> - consider the true impact of communication from a more philosophical point of view.

2.1 Communication—A Universal Concept

Viewed as a word, *communication* is certainly a handful.

Firstly, it is way too long and bulky with its five syllables and thirteen letters. Handling it in a text is like carrying a 10-foot wooden plank inside an office; it constantly creates practical problems connected to text flow, layout, and typography.

Secondly, as a term, it is vague and inexact, and, annoyingly often, it is used without being properly defined. When using words, terms and expressions, navigating in the dark is never a wise thing to do (we'll come back to that in Sect. 3.1). Therefore, let's take a fresh look—in Fig. 2.1—at the dictionary entry for *"communication"* to make sense of this unruly concept.

Communication

From Latin *communicare* meaning "to make common", which stems from *communis*, "shared"

Fig. 2.1 The word *communication* itself has been used in the English language at least since the sixteenth century. The etymology is quite straightforward: it is derived from the Latin word *communis* 'common' via *communicare* 'to make something common; to share.' Related words are *communion, commune,* and *communist*

The best definition we've found so far was actually included[1] in an earlier version of the Wikipedia article about communication.[2] Why this one? Because it shows the multi-branched, universal nature of communication! Here it is:

'the activity of conveying information through the exchange of ideas, feelings, intentions, attitudes, expectations, perceptions or commands, as by speech, gestures, writings, behavior and possibly by other means such as electromagnetic, chemical or physical phenomena. It is the meaningful exchange of information between two or more participants (machines, organisms or their parts).'

> **Reflection II.I**
> To get a demonstration of how the concept of communication spans an enormous range of human activities, you can do a little experiment: visit an online bookstore, type the word "communication" in the search field, and study what you get.
>
> Most likely, a long list of book titles will appear, representing the most diverse subjects: psychology, leadership, visual design, networking technology, and so on. Based on this, it is no bold statement to claim that communication is involved in everything we humans do.

---

[1] Unfortunately, *Anonymous* decided to replace it with one that is much vaguer.
[2] Wikipedia Contributors (2019, April 22).

2.2 Modes of Communication

2.2.1 Suggesting a Seven Stage Model

Based on the Wikipedia definition above, we note that communication can be anything from the most subtle, personal thing (for example a quick, telling glance from a fellow person) to a huge, noisy techno-social system (for example the Oscars awards gala).

To get a historical overview of what the concept "human communication" has actually included through the ages, we suggest a model which forms a timeline, divided into seven stages. The model is a bit schematic and suffers from an apparent Eurocentric perspective. But please bear with us: we are all science people who can handle models to help us think, understand and remember—right? (Fig. 2.2).

Prehistory: The evolutionary and early cultural stage
During the reading of Yuval Noah Harari's popular book *Sapiens*,[3] it becomes apparent that communication has played a central part in our species' evolutionary success. The exact point at which biological behaviors evolved into something that may be

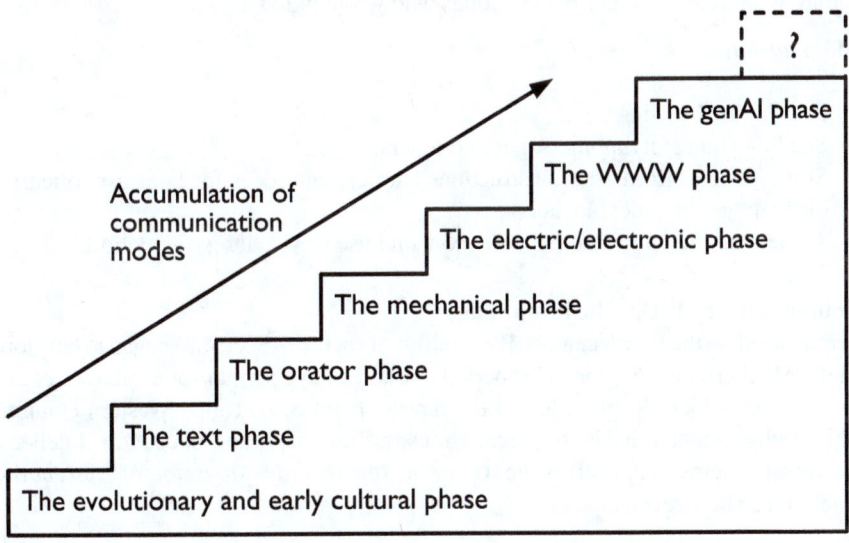

Fig. 2.2 A summary of the seven stage model of communication

---

[3] Harari (2019).

called a cultural pattern is hard to define. If we arbitrarily use the dawn of agriculture and permanent settlements as the starting point for the timeline, it shouldn't be a daring statement to claim that humans have "always" engaged in the following types of communication:

- Understanding and using touch, sounds, body language, gestures and facial expressions.
- Understanding and performing language.
- Understanding, interpreting, inventing, and re-enacting or performing stories.
- Understanding and creating metaphors.
- Understanding and creating visual representations, including those of people, animals, events, objects, locations, and ideas, as well as semiotic communication in the form of signs, symbols, banners, colors, et cetera.
- Understanding, interpreting, inventing, and performing (oral) literature.
- Interpreting, constructing, and performing rites and ceremonies.
- Appreciating and creating music.

3000 B.C.: The text stage

With the advent of written communication systems (Egypt, Mesopotamia, China and others), many limitations of human communication were swept away—not least temporal and spatial constraints. This meant that data, information, know-how and stories now could be stored far more effectively outside the human brain, beyond just visual representations. Written knowledge could also be used, interpreted, and retold by individuals other than the ones who wrote it down.

Main advances:

- Reading and writing text.
- Sending and receiving messages and letters.
- Storing data, information, instructions, stories, and ideas for later use, often in different geographical locations.
- Advancing the understanding, creation, and use of visual representation.

Fourth century B.C.: The orator stage

From the 6th to the fourth century B.C., the art of rhetoric (Sect. 4.3) began to develop in the Mediterranean region. This period marked a huge leap in some areas of communication, which deeply affected the development of so-called Western culture: philosophy, science, business, negotiations, politics, law and education. Intellectual hotspots emerged, such as the *Agora* in Athens or the *Mouseion* of Alexandria (including The Great Library).

Main advances:

- Planning and performing strategic communication, with the aim to persuade target groups and achieve goals.

- Using dialectic and logic for philosophical and scientific exploration, argumentation, deduction, and proof.
- Refining a range of communication channels and art genres.

Fifteenth century: The mechanical stage

For almost two thousand years, no radical changes occurred in the way people communicated. Different civilizations largely confined their activities to their parts of the globe, the literate among them relying on pen and paper—note, for instance, the Christian civilization in Europe, where generations of monks and scholars spent their time meticulously copying hand-written books. Then, a 1450s invention changed what we call Western culture forever: the European implementation of printing press technology.[4] The first Bibles printed by Johannes Gutenberg in Mainz, Germany, started the Printing Revolution, and marked a new era of communication technologies based on wood, metal, and mechanics.

Main advances:

- Producing books, newspapers, leaflets, art reproduction, and other printed materials in many copies, at a faster rate and a lower cost.
- Producing large-scale visual boards and posters.
- Projecting images (e.g., the *laterna magica*[5]).
- Refining a range of communication channels and art genres.

Nineteen–Twentieth century: The electric/electronic stage

Following the *Age of Enlightenment* of the eighteenth century, innovation accelerated. Early scientific methods paved the way for, firstly, steam technology and refined mechanics and, secondly, the breakthrough of electricity—which in turn developed into transistor-based electronics, followed by integrated circuits of semiconductor material ("computer chips"). In parallel, other technologies emerged, like photography, optics and radio technology. All this made communication easier, faster and created new types of media.

Main advances:

- Using cable-based technology to—at first—transmit text (the telegraph with its Morse code), and—some decades later—to transmit audio (the telephone).
- Using radio-based communication, first to carry text (in Morse), later sound (radio) and moving images (TV).

---

[4] Here we should raise a red flag of Eurocentricity and establish that movable type is actually a Chinese invention from the eleventh century. Johannes Gutenberg's achievement is not a proper "first of its kind"; instead it could be compared with the Macintosh computer in 1984, representing an integration and streamlining of a set of useful inventions into a powerful main concept.
[5] A kerosene-illuminated forefather of the conference room projector.

- Using silver-based technology, enhanced with different types of lenses, to produce images of actual objects, places, and events (photography), which eventually developed into moving images (film).
- Refining a range of communication channels and art genres.

Late twentieth century: The WWW stage
Writers and readers of this book have all been embraced by the enormous changes set about by internet technology—a Cold War infrastructure for digital signal communication that became the foundation for increasingly powerful protocols and applications. Some of us saw it emerge, others were born as natives of this Brave New World. It was the advent of the World Wide Web—during the first part of the nineties—which made this advanced technology truly useful for a wider population.

Main advances:

- Using hypertext technology to distribute information and offer a platform for different kinds of interactive technology.
- Using social media.
- Using online video conferencing.
- Democratizing publishing, allowing anyone and everyone to communicate their facts, emotions and opinions to the world.
- Refining a range of communication channels and art genres.
- Storing enormous amounts of information in server parks.

Early twenty-first century: The AI stage
Wondrously enough, humanity entered a new phase after the contract for the book you're reading right now was written up. On 30 November, 2022, ChatGPT was made available to the public, and it is no longer always possible to distinguish text, speech, image or video made by a machine from those made by a human. Globally, researchers, intellectuals, media professionals and writers are trying to get their heads around the disruption that is starting to happen. Where will this take us?

Reflection II.II
Tape two blank pages together at the short edges and draw a timeline from ancient times to the year 2200. Place it on the table before you and start doodling and scribbling. What do you remember about the great changes in history? Which phases are most interesting to you? What have we omitted in the seven-stage model? What do you think will happen in the two centuries ahead?

2.2.2 Learning from the Seven-Stage Model

Let's reflect on this timeline-based overview to see what it may tell us.

- **Communication methods from all seven stages are constantly in use—here and now, and in parallel.**

  Once communication technologies or genres prove themselves useful or influential enough, many of them persist indefinitely. This makes the seven-stage model cumulative, meaning that elements from earlier stages are retained in later ones.

  Interestingly, the earlier, more "primitive" stages may play a more important role than is first apparent.

  Example 1: During a discussion at a scientific conference, the persuasive power of the argument may be heavily influenced by social power manifestations, where body language and tone of voice play a central role. To put it more bluntly: sometimes, threatened seniors bully gifted juniors into submission.

  Example 2: Writing skills, wit, and general eloquence are a requisite for success in social media; the same is true for illustration and photography. Hence, there is little doubt that pre-digital, multi-gifted artists like Virginia Woolf, Paul Klee, and Man Ray, would have thrived in the social media environment.

- **The tools of the first stages are often the most powerful ones.**

  Effective communicators often do away with the sophistication of our civilization and technology. Raw, simple acts can be so much more potent than all the refinement our modern channels offer: banging your fist on the table, speaking with a trembling voice while barely holding back tears, tearing a paper apart, or literally drawing a line in the sand to separate you from the person you're talking to.

- **There are, most likely, unknown communication tools and channels ahead of us.**

  As the communication toolbox of *Homo sapiens* has been growing continuously, it is not unreasonable to assume that it will keep growing.

So what's around the corner? The first thing that comes to our mind are the VR-environments where teachers and students enter the same interactive simulation and walk around in their equations while studying mathematics. We also believe that a wide range of hybrid formats are taking shape. The most apparent is that computer games and generative artificial intelligence are approaching the movie genre, creating immersive and interactive experiences. But it can be as simple as using poetry for communicating science, or using comics in corporate training.

And after that? Well, you tell us!

> **Reflection II.III**
>
> What unknown communication tools and channels do you think people in the future will use? Have you seen some interesting concepts in, for example, science fiction novels or movies? How will these tools change social, cultural, and economic life?

2.3 Three Observations Regarding Communication

The universal nature of communication has, of course, some universal implications. Below, we will discuss the following statements:

- It is impossible *not* to communicate
- To communicate is to take part in the business of the world
- Communication is the core matter of any human activity.

It is impossible *not* to communicate

In 1967, communication theorists Paul Watzlawick, Janet Beavin Bavelas and Don D. Jackson published the book *Pragmatics of Human Communication*[6] where they defined five axioms of communication. While four of them are very dry and theoretical, the first has become a classic:

"*One cannot not communicate.*"

Think about it for a while: staying silent often conveys a statement, walking out of a room may send a very clear message, not showing up makes everybody wonder … and so on. Even the blank space between this paragraph and the next is a part of our writing (Fig. 2.3).

Fig. 2.3 If you don't get the reference here, please look up "La Trahison des Images"; it refers to a famous painting by the Belgian artist René Magritte (1898–1967)

---

[6] Watzlawick et al. (1967).

2.3 Three Observations Regarding Communication

The lesson here is this: even when you abstain from active communication you should consider the people on the receiving end. For example, if you fail to reply promptly to an email that is essential for the receiver but peripheral for you, you may cause much worry. And although you're free to skip the details of your latest research when giving a presentation—just because you want to focus on an earlier project—it can lead the audience to quietly draw their own conclusions.[7]

To communicate is to take part in the business of the world
Let's take inspiration from the British language philosopher J. L. Austin (1911–1960) and his famous book *How to Do Things with Words*.[8] Communication is not merely about stirring up things in an abstract, metaphysical universe; it is about promoting action (we will revisit this idea in Sect. 3.2). The result may be seen right away, or may require a ten year wait—what's important here is that your communication activities prompt some form of action. In a scientific setting, these actions may be:

- transferring information or instructions;
- creating understanding;
- motivating people and establishing trust;
- bringing members of a group closer to one another;
- convincing an opponent;
- selling an idea or a solution;
- influencing decisions.

We'll come back to this bullet list and have a closer look at it in Sect. 4.4.1.

Communication is the core matter of any human activity
Most importantly, the seven-stage model proves a central point we want to make with this book:
Communication is the core matter of any human activity.[9]

---

[7] Speaking about people trying not to communicate, here's a general observation we've made at international border controls. Where people are anxious that they will get hassled by officials, you can spot a special kind of facial expression: a totally blank look, giving away no human emotion. The sterner the border guards, the more this look spreads among the people standing in line. What these people are trying to express is this: "Nothing to see here, I am not the person you're looking for to stall, search, or harass." A message conveyed just through these blank looks!

[8] Austin (1962).

[9] Of course it is possible to create a theoretical example of a Robinson Crusoe, marooned on an island, who is solving problems and surviving without communicating with either humans or animals. However, these persons are most likely very few and they certainly don't have to feel that they are the target group of this book (remember that in the movie *Cast Away*, Tom Hanks' urge to communicate with another soul became so strong that he created Wilson the Volleyball to get some company!).

Whatever you set out to do, success is impossible without excellent communication—whether it's passing your exam, learning to ski, inventing the next disruptive lab methodology, forming a rock band, explaining the mechanism behind a rare kind of tumor, having a happy marriage, creating a breakthrough vaccine, building a summer house, or winning the Nobel Prize ... the list is endless.

In fact, top-notch communication sometimes represents the best that life has to offer: an unforgettable dinner conversation with witty friends, an awe-inspiring concert by a creative band, a thought-provoking text by a brilliant fiction writer. Conversely, one of the most severe punishments in the penal system of democratic states is solitary confinement.

> **Knowing a little about a lot of things—why general knowledge and *bildung* are so important**
>
> General knowledge is extremely important for a professional communicator. To explain why, let's start with the concept of *Bildung* which plays an important role in German cultural tradition.
>
> The word *Bildung* itself is hard to translate into a single English word; it could be seen as a merge of *education* and *formation*. The idea is that we should all cultivate our own minds and that learning, reading, observation, and intellectual interaction with others lead to personal and cultural maturation. Essentially, the more you learn, reflect and gain insight, the better person you become.
>
> In literary studies, the term *Bildungsroman* refers to a type of novel that explores how a protagonist develops as a person from a moral and psychological point of view. "Coming of age" is a similar concept, where the reader follows a young person's transition from childhood to adult life. Famous examples are *Demian* by Hermann Hesse, *The Catcher in the Rye* by J. D. Salinger, and *The Kite Runner* by Khaled Hosseini.
>
> The common expression for 'knowing a little about a lot of things', accordingly, has some gravity in Germanic languages: *Allgemeinbildung* (German), *allmänbildning* (Swedish), and so on. In the English context of this book, we must stick with the expression *general knowledge*, which appears somewhat bland in comparison.
>
> Nevertheless, general knowledge is extremely important for a professional communicator. It helps the author find intriguing connotations and connections. It helps the journalist see through cover-ups and ask more pertinent questions. It helps the editor spot errors on the pages of books and magazines. It helps the copywriter and art director come up with more creative ideas. Overall, research suggests that general knowledge actually makes your mind more agile, productive and far-reaching.

2.3 Three Observations Regarding Communication

Apart from increasing your intellectual ability, general knowledge also helps communicators of all kinds connect with their target group. The principle is easy to put into practice. As Dale Carnegie once pointed out in his famous book *How to Win Friends and Influence People*,[10] most people will see you as a considerate person if you show interest in what's important in their life, and—in response—open up for communication. Therefore, knowing a little about world geography, countries, capitals, flags, languages, history, sports, et cetera can help you connect quickly with the stranger in front of you.

General knowledge also helps you to be kind and friendly to people without becoming too personal. Just like the weather, geography and history can serve as platforms for so-called 'phatic communion'—communication that serves to form a social relationship between individuals, rather than just conveying information. For example, when people are allowed to talk about their own culture or hometown, it may help repair their self-esteem in a situation where they feel far from home and out of place.

Reflection II.IV
Next time you're going to meet a new colleague, take some time to read the Wikipedia articles about their country, hometown and university. What has been said recently about these places in the news?

After your first day together, assess the interaction and conversation you had. Was this background knowledge helpful? Did it lead to new insights or open any new perspectives?

---

[10] This book has often been mocked as it is overly optimistic and only based on anecdotes and the author's personal experience. Still, his "six ways to make people like you" can definitely be defended from a scientific perspective, see for example https://www.bakadesuyo.com/2013/01/truth-dale-carnegies-how-win-friends-influence-people/. For a large part, Carnegie is just putting basic rhetorical principles to the test.

References

Austin, J. L. (1962). *How to do things with words.* Oxford University Press.
Carnegie, D. (1936). *How to win friends and influence people.* Simon & Schuster.
Harari, Y. N. (2019). *Sapiens: A brief history of humankind.* Random House.
Watzlawick, P., Bavelas, J. B., Jackson, D. D., & Norton, W. W. (1967). *Pragmatics of human communication: a study of interactional patterns, pathologies, and paradoxes.* W.W. Norton & Company.
Wikipedia Contributors. (2019, April 22). *Communication.* Wikipedia; Wikimedia Foundation. Retrieved August 15, 2024, from https://en.wikipedia.org/wiki/Communication

Several Handy Theoretical Models for Practical Communication

A Set of Shortcuts in the Maze of Communication Theories

> **What you will learn from this chapter**
>
> Communication has inspired an entire, very rich, research field. While going through all of it might be somewhat overwhelming, we would like to point you to a few theories that can be easily applied in your everyday communication activities.
>
> In this chapter, you will:
>
> - reflect on the language (concepts, terminology, jargon) that scientists use on a daily basis—and what purposes it serves;
> - be introduced to the Shannon-Weaver model of communication—our favorite model;
> - learn about other useful communication models from a range of disciplines.

3.1 Introducing a Bit of Theory

In Chap. 2, we noted that the concept of communication may appear surprisingly hard to grasp as soon as one tries to define it or seek agreement on how to approach it. Consequently, entering the discourse of what armies of scholars have to say about communication is like entering a never-ending maze of mind-spinning theory, explanations, and models.

To avoid getting lost, we will try to simplify this whole field of study; our aim is to provide some handy shortcuts from the maze that can be used practically—here and now. Accordingly, we have picked some 20th-century models or principles in Sect. 3.2. They are chosen to be easily applicable in the real world when you approach different scenarios or tasks—either for analyzing, planning or performing an activity.

But before we move into these topics, let's discuss definitions and terminology.

3.1.1 Let's Define the Concepts We Use

Educated people of the world share a bad habit: they tend to discuss things without ensuring they have an agreement on definitions.[1] No area is excluded; the same phenomenon can be seen in academia, industry, media and politics.

Example: What does "juvenile" mean?
When one of us (Joanna) was running a Three Minute Thesis workshop with doctoral students from a range of disciplines, she asked them to partner up with someone from a different department and swap outlines of their doctoral projects. One pair—a psychologist and a lawyer—were visibly confused. It turned out that one of them used the term "juvenile" in their outline, which means something very specific in psychology—and something equally specific (but different!) in law. To make matters more confusing, in ecology it means something different as well. Think about it: "juvenile" may refer to immature behavior, an offender who is too young to be prosecuted as an adult, an animal that hasn't reached reproductive maturity—or water that's derived from the earth and reaches the surface for the first time. Mind boggles.

Sometimes it's not the jargon that we need to be wary of, but the terms that seem so obvious that we don't bother to define them. Therefore, our first piece of advice in this chapter—which is applicable to any intellectual activity—is to make sure that the core concepts are clarified. As a matter of fact, it is a great start for any project—or any conversation!

3.1.2 Let's also Agree on the Terminology

In most cases, it is not only necessary to agree on central concepts—we also have to cultivate a specialized language with our peers. *Terminology* is the word used to describe a collection of special words—or alternative uses of common words—specific to a particular field.

This is something very typically human and very universal. Throughout history, whenever people have formed groups, they have also created languages tailored to describe and manage their specific contexts. Depending on factors like social structure, genetic heritage, migration waves, and geographical distribution, we categorize these languages in different ways: *language groups* are divided into *languages*, which are divided into *dialects*. To describe even smaller subsets, we can introduce the term *microlect* which can even refer to the language used between two individuals (for example, a married couple using silly expressions from their early, romantic days together).

As soon as a group of people becomes professionally specialized, they need to create their own microlect to be able to work effectively without misunderstandings. For instance, nineteenth century sailors needed the word *mizzenmast*, whereas

---

[1] Davie (2012, August 15).

midwives of today use the expression *mucous plug*, and modern engineers may talk about *shearing strain*; you can most likely rattle off dozens of such words in your sleep from your own professional domain. Overall, the number of work-related expressions across different professions is much greater than the number of words we use in the common language.

When a complete, specialized terminology is developed for a particular field or area of activity, it is called *jargon*. Most of it consists of a set of *terms*, which Wikipedia defines as words or expressions "that in specific contexts are given specific meanings."[2] The meaning of these terms "may deviate from the meanings the same words have in other contexts and in everyday language." For example, in obstetric care, the occupational term *spontaneous abortion* should be avoided when healthcare staff communicate with patients. The reason is that the word *abortion* is commonly associated with elective termination. Instead, the layperson's expression *miscarriage* should instead be used to avoid misunderstanding.

Jargon is very interesting to study from a sociopsychological perspective, as it easily evolves into a *sociolect* (that is, the language of a certain group). It can be used to keep a community together—which is nice—but also to keep outsiders at bay—which is not always that nice. When phrases, expressions, and words are used to distinguish one group of people from another, they are known as *shibboleths*.[3] Unfortunately, sociolects can also be the perfect breeding environment for *buzzwords*; see Fig. 3.1.

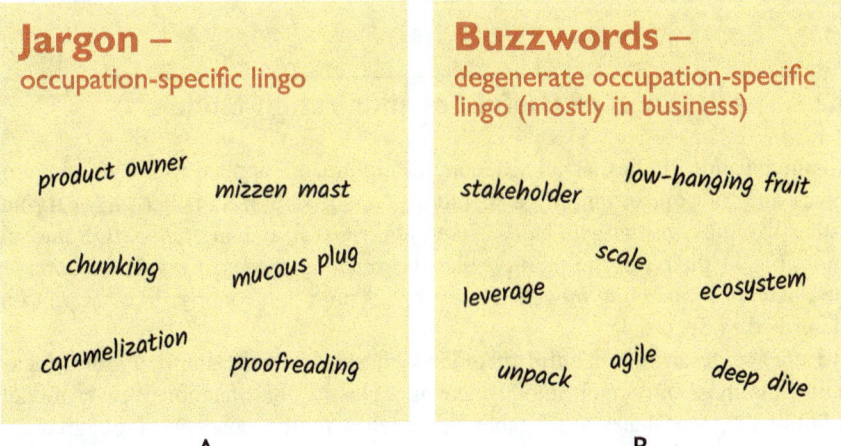

Fig. 3.1 Every profession has its own lingo (**a**). It makes communication fast, while avoiding misunderstanding. Unfortunately, some jargon words mutate into buzzwords—words or phrases that are fashionable (**b**). They signal that you are part of the game, but really only serve as kind of glorified filler words

---

[2] Wikipedia Contributors. (2021, January 6).
[3] This comes from a Biblical story where two tribes who were in conflict spoke the same language but tended to pronounce certain sounds differently; in the story, the Hebrew word *shibboleth* was used as a password to tell friend from foe.

> **Reflection III.I**
> Have a look at the Fig. 3.1a. Can you identify the professional areas where the words are used?
>
> Write down three words that you use in your field. How would you explain them to a lay person?

How, then, should we, as communicators of science, approach terminology?

- It is useful for us to be aware of the jargon we use with our colleagues. We should let the terms be effective tools for peer-to-peer communication, rather than shibboleths that distance us from people outside our tribe. Using jargon to make others feel inferior is a sign of a very immature mind.
- When communicating with lay people, we ought to examine the professional terminology used in our field and adopt one of three approaches:

    a. **Replace** unnecessarily complicated terms with plain English
    b. **Explain** terms that are difficult to replace, clarifying them for the reader
    c. **Teach** terms that are central to the subject (for example, *vesicles* when we're talking cell biology, or *alpha particles* when we discuss nuclear safety), defining them clearly so that they can be understood and memorized by our audience.

3.2 A Handful of Useful Communication Models

Communication studies, as an academic discipline, are applied to a wide range of topics and, of course, employ a plethora of models. Some of them have formal names like the *transactional model, interaction model,* and *micromodels.* Some are intriguing in their impenetrability, like the *semiotic model of communication* or the *meaning of meaning model.* Others sound more welcoming, like *storytelling* (described in Sect. 6.4).

Here we describe a handful of models chosen for their simplicity and practicality; the most important among them is a classic: the Shannon-Weaver model. Additionally, we share some other approaches to the concept of communication, all of which convey a central wisdom that can be turned into guidelines, rules-of-thumb and practical advice. These come in handy when you—on the one hand—observe and analyze communication in general, and—on the other hand—plan, prepare, and execute your specific communication activities.

Main model

- **The Shannon-Weaver model**

  Central wisdom: clarifies the relation between you, your message, your receiver, and the context of the communication activity.

Auxiliary models

- **The Lasswell formula**

  Central wisdom: the mnemonic *"Who says what, via which channel, to whom, with what effect?"*

- **The rhetorical principle of J. L. Austin: "How to do things with words"**

  Central wisdom: to communicate is to take action.

- **The user stories employed in software development**

  Central wisdom: clarifies the relationship between you, your message and material, your receiver, and the goal of the communication activity.

- **Business problem model**

  Central wisdom: the problem-oriented format offers clear and solution-oriented communication, closely related to storytelling.

3.2.1 The Shannon-Weaver Model—The Boss of All Communication Models

When browsing through textbooks and websites about communication, you will most likely encounter the Shannon-Weaver model. Let's take a closer look at it.

In the model illustrated by Fig. 3.2, the **sender** transmits a **message**—which is **encoded** in some way—through a **channel**. During transmission, the message is distorted (to varying degrees) by **noise**. The message is then **decoded** and reaches the **receiver**.

Shannon-Weaver model of communication

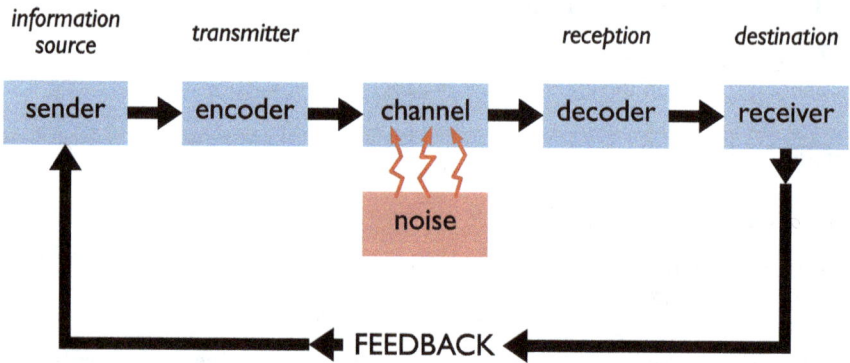

Fig. 3.2 A summary of the Shannon-Weaver model of communication

Example: I [SENDER] write a secret classroom note [CHANNEL] to you with a joke in hand-written English [ENCODED MESSAGE]. My pencil is unsharpened, making the writing somewhat difficult to read [NOISE]. I throw the note in the form of a paper ball to you [RECEIVER] when the teacher turns his back to us. You read it with some difficulty, understand my joke [DECODED MESSAGE] and give me a smile and a thumbs up [FEEDBACK].

Claude Shannon, a mathematician and engineer from the University of Michigan, first came up with the model, and then popularized it in the book *The Mathematical Theory of Communication*[4] with help from Warren Weaver (also a mathematician). If you're into engineering sciences, the model will probably remind you of the feedback loop diagrams used in control theory.

On its own, without guidance or examples, this model may seem to offer little practical help. But we have meddled with it, removed some stuff and added other stuff, all with the ambition to make it practical and hands-on. And this is what we came up with: Five Universal Guidelines for effective communication activities. We will have a closer look at these guidelines in Chap. 6, but here is their essence (refer to Fig. 3.3 to see which parts of the model they relate to).

---

[4] Shannon and Weaver (1998).

Five universal guidelines for effective communication

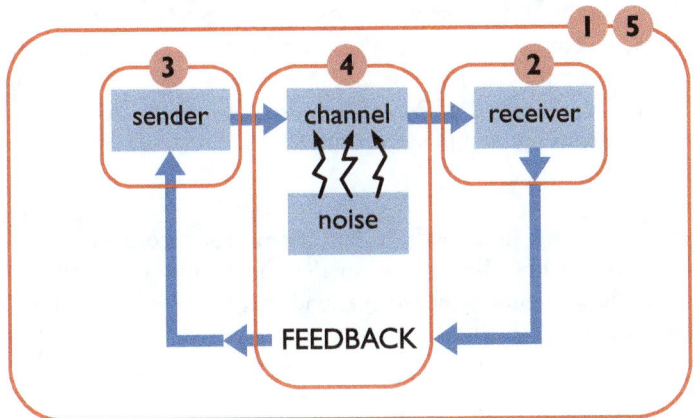

Fig. 3.3 Five universal guidelines stemming from the Shannon-Weaver model

Guideline 1: Define your message and goals.
At the center of the Shannon-Weaver model is the message being conveyed. But we also introduce a meta-level: what is the desired outcome of the communication process?

Guideline 2: Know your target group.
To ensure effective communication, scrutinize the receiver in the S-W Model; this process is normally referred to as a target group analysis.

Guideline 3: Know yourself.
Get a solid grip on who the sender is in the S-W Model—typically, that's you!

Guideline 4: Understand the limitations at hand.
To control the communication process represented by the S-W Model, understand all its details—the nature of the channel, the noise, etc. this includes knowing what's possible, and what isn't.

Guideline 5: Seek inspiration in all types of communication.
Skilled communicators never settle for mastering only one specific setup of the S-W Model. Instead, they explore what happens when contexts or parameters change. It may be especially effective to study the use of different channels—what works and what doesn't, when, and why.

> **Effective versus efficient communication**
> Please note that in this book, we discuss the concept of "effective" rather than "efficient" communication. Let's have a look at their definitions:
>
> - effective: 'successful in producing a desired or intended result'
> - efficient: 'achieving maximum productivity with minimum wasted effort or expense'.
>
> The overarching goal is to achieve the desired outcome, not to earn style points for flawless organization and spotless perfection while getting there! As the management guru Peter Drucker once put it: "Efficiency is concerned with doing things right. Effectiveness is doing the right things."[5]

3.2.2 Auxiliary Models

To complement the Shannon-Weaver, we add other models (tools, approaches, or whatever you want to call them) that can help you analyze, prepare and execute your communication activities.

The Lasswell formula: "Who says what ..."
In 1948, Yale professor Harold Lasswell published a highly influential article called *The Structure and Function of Communication in Society*.[6] He concluded that any act of communication can be described in the following way:

Who says **what** in **which channel** to **whom** with **what effect**?

Ten years later, in his article "An extension of the 'Lasswell formula,'"[7] scholar Richard Braddock added two more elements to the above definition:

Who says **what** to **whom** under **what circumstances** through **what medium** for **what purpose** with **what effect**?

These models are useful for information evaluation, not least in historical and political studies.

Example: *Traditionally, English king Richard III (1452–85) is portrayed as a historical villain. But was he? Let us use Braddock's formula, with FOR WHAT PURPOSE? as the "unknown variable":*

Through **his famous play, Shakespeare** *tells* **his sixteenth/seventeenth century-audience** *[FOR WHAT PURPOSE?] that* **Richard III was an evil ruler**, *with the effect* **that it has become the general opinion**.

---

[5] A common misquotation circulated on the web is this: "Management is doing things right; leadership is doing the right things."
[6] Lasswell (1948).
[7] Braddock (1958).

After reading this, things don't seem so obvious anymore. What would Shakespeare's true purpose be? A likely reason is that using a dead ruler with an already poor reputation and turning him into a full-blown Machiavellian monster is a very effective tool for a playwright, since the audience will quickly understand the good-vs.-bad structure of the story.

So, although the discussion above doesn't settle the Richard III question, it is apparent that it can help us approach the problem in a structured way.

To sum up, the Lasswell and Braddock formulas are useful reminders when attempting to interpret the world around you—especially when you want to discover the drivers and motivations at the core of communication activities.

The Rhetorical Principle of J. L. Austin: "How To Do Things With Words"

In 1955, the Oxford philosopher of language J. L. Austin delivered a famous lecture that became his intellectual legacy.[8] In it, he introduced what would later be known as *the theory of speech acts*. What's interesting for us is his emphasis that language could be used not only to *assert* things, but to *do* things. Or, as Wikipedia puts it: "Austin's work ultimately suggests that all speech and all utterance is the doing of something with words and signs". Examples of speech acts include announcing, answering, apologizing, complaining, confirming, congratulating, inviting, ordering, proclaiming, promising, refusing, requesting, and warning. Since we often express certain desires or goals in our communication, or actively seek something from our audience, it is very hard to define which verbal messages are speech acts and which are not.

How can this insight be useful for scientists who are less interested in language philosophy, and more interested in using communication to boost their career? Well, they should recognize that anyone who *is communicating* is in many situations also *taking action, and changing things*. Language and, in a wider sense, communication does not belong to the field of metaphysics. Instead, it should be seen as something that influences and alters things—not only in the material world, but also in the professional, social, cultural, and political worlds. To put it bluntly, words have consequences: emotional, medical, financial, and so on.

We could sum it up by paraphrasing the famous quote by the French philosopher René Descartes (1596–1650) "I think, therefore I am" (*Cogito ergo sum*) in this way:

I communicate, therefore I take action

(*Communicare ergo ago*).

The User Stories Employed in Software Development

User Stories, a tool used by software developers to create effective and end-user-oriented software, may seem an odd communication model, but bear with us. The concept is extremely simple: a short and informal description in plain language of what the *end user* (that is, not the developer!) wants to do in a computer program to

---

[8] Austin (1975).

achieve something valuable. These little instructions have been around for about 20 years and—since software developers love tweaking and modifying their methods and ways of working—can be found in different varieties; three of them appear here:

> **As a <role> I can <capability>, so that <receive benefit>**
> Example: As the owner of a CRM database, I can sort all the entries by surname so that I can export alphabetic address lists.
> **In order to <receive benefit> as a <role>, I can <goal/desire>**
> Example: In order to export alphabetic address lists as the owner of the database, I can sort the entries by surname.
> **As <who> <when> <where>, I <want> because <why>**
> Example: As the owner of the database, during office hours, at my laptop, I want to sort the entries by surname because I need alphabetical address lists.

Now, let's use these language structures to plan some communication.

> As a presenter, I can boost my credibility by being enthusiastic and well-prepared, so that important people in the audience find me trustworthy and reach out to seek interaction.

or:

> In order to stand out among the candidates as someone being well-matched with the job description and an interesting future colleague, as the job applicant, I can write a cover letter that explains how my qualifications match the requirements and also conveys a portrait of a committed professional.

or:

> As a fresh postdoc within a year of my Ph.D. at my new lab, I will write an ambitious review article about the Sonic Hedgehog (SHH) protein signaling in tumor cells because I want to establish myself as a serious player in the field.

Business Problem Model

The model used for communicating problem-solving in the business world is based around the following premise: the short-term solution is not the end of the story; for the solution to prove itself feasible, there must be a positive long-term outcome.

$$\text{Problem} \Rightarrow \text{Solution} \Rightarrow \text{Outcome}$$

To use this model for communication, a step can be added at the start to create some context.

$$\text{Situation} \Rightarrow \text{Problem} \Rightarrow \text{Solution} \Rightarrow \text{Outcome}$$

What we have here is a storytelling structure! It is very useful for making presentations about ongoing research more interesting, creating a hero's quest where you patiently overcome setbacks in the lab to reach an exciting and happy ending (read more about storytelling in Sect. 6.4).

Example: HPV Prevention
Problem: *Virus-mediated cervical cancer kills hundreds of thousands of women worldwide yearly.*
Solution: *Human papillomavirus infection was identified as the cause in 90% of the cases. This paved the way for effective vaccines now included in routine vaccinations in over 70 countries.*
Outcome: *Although still early in the process (as cancer is mostly associated with older age), several countries have reported a decreasing incidence of cervical cancer in HPV-vaccinated populations. Many clinicians hope that HPV-mediated cancers can be eradicated in the future.*

Tips from corporate life:

1. **Adapt to your audience**: Understand and respect cultural differences in communication. Adjust your style to suit the preferences and norms of your audience, whether it involves assertiveness, formality, or giving feedback.
2. **Start with the key point**: In a busy corporate setting, be direct in your communication. Clearly state your purpose, whether it's seeking help, an opinion, or a decision, to save time and maintain efficiency.
3. **Emphasize purpose in presentations**: When presenting, focus on the 'why' behind your presentation, not just the 'what'. Aim to drive the conversation forward for actionable outcomes, rather than just sharing information.

Acknowledgements Thank you to Martina Björk, Lund University, for the translation of the paraphrased Descartes quote.

References

Austin, J. L. (1975). *How to do things with words*. Harvard University Press.
Braddock, R. (1958). An extension of the "Lasswell Formula." *Journal of Communication, 8*(2), 88–93. https://doi.org/10.1111/j.1460-2466.1958.tb01138.x
Davie, G. (2012, August 15). *How to define your concept a.k.a. concept explication [Part 1]*. Mass Communication Theory. Retrieved July 15, 2022, from https://masscommtheory.com/2012/08/15/how-to-define-your-concept-a-k-a-concept-explication-part-1

Lasswell, H. (1948). The structure and function of communication in society. In L. Bryson (Ed.), *The communication of ideas.* Institute for Religious and Social Studies.

Shannon, C. E., & Weaver, W. (1998). *The mathematical theory of communication.* University of Illinois Press.

Wikipedia Contributors. (2021, January 6). *Jargon.* Wikipedia. Retrieved January 22, 2021, from https://en.wikipedia.org/wiki/Jargon

Understanding the Universal Usefulness of Rhetoric

Why Clever Communication Strategies by Dead Guys Are Still Relevant Today

> **What you will learn from this chapter**
> In this chapter, we introduce rhetoric in both a historical and contemporary context. We also provide some useful rhetorical principles and tips to apply in your communications activity, such as
>
> - the Five Ws, so popular in journalism today,
> - the universal Aristotelian triad of ethos, pathos and logos,
> - the ubiquitous Rule of Three (once you know about it, you'll notice it everywhere).

4.1 What is Rhetoric All About?

What do you associate with the word "rhetoric"?

- Bombastic but empty communication, as in "this political announcement is just empty rhetoric"?
- A flashback to high school history lessons about stern guys in togas, waving concepts like *dispositio* and *vir bonus*, while insisting that Carthage must be destroyed?
- An experience-based way of approaching the act of communicating, whether it be analysis, planning, or performance—an old, yet vital art studied and applied by scholars, teachers, and PR people alike?

This book, of course, opts for the third standpoint. To us, rhetoric represents the most comprehensive and battle-proven toolbox of communication know-how. Here we aim to provide you with the essential essence of rhetoric.

First of all, regarding rhetoric as simply "the art of speech" is very limiting. Today, the concept goes far beyond public speaking,[1,2] and can be defined in several ways. Here are some examples:

- *The art of persuasion.*
- *The art of effective or persuasive speaking or writing.*
- *A way of doing things with words.*
- *"The faculty of observing in any given case the available means of persuasion"* (this definition comes from Aristotle himself—read more in the fact box below).
- *Communicating effectively based on an understanding of human nature* (this definition comes from the writers of this book).

In academia, rhetoric is a field of study rather than a science. It encompasses a body of practical and theoretical knowledge accumulated over more than two millennia. It has had its ups and downs, but today it is alive and well as a branch of learning studied and taught by universities all over the world. Besides being defined as an academic discipline, it is often described as an *art*, much like *the art of war* or *the art of cooking*—a blend of intellectual and practical skills.

Some communicators want to set an ethical criterion for rhetoric, suggesting it should only count as "true" rhetoric if the purpose is good. We believe this way of thinking is constraining because studying masterful communicators from "The Dark Side" can help us fight fraud, manipulation, and bullying—perhaps even gain a clearer view of how human evil works, or at least the mechanisms of how "good people do bad things."

It is important to understand that rhetoric is not about who's actually right from a factual point of view, but rather who has the power to convince an audience of what is "true." This can be a very distressing thought for science people with a sincere passion for evidence-based truth. However, in a world of *fake news* and alarming statistics regarding scientific misconduct, it is essential to understand why people choose to embrace some claims and reject others when it comes to how reality is described.

4.2 Rhetoric is Everywhere

The historical impact of rhetoric is hard to exaggerate. Rulers and politicians throughout history have been taught and trained in the art and have applied it to change the course of history innumerable times. But what's even more important is that concepts, methods, and structures from rhetoric can be found around

---

[1] Drout (2006).
[2] Leith (2011).

4.2 Rhetoric is Everywhere

us, everywhere and all of the time—academia included. For centuries, scholars, teachers, and students of the Western university system have been like fish swimming in an aquarium of rhetoric. Today, many seem unaware of this fact—probably because it's part of their intellectual environment.

In fact, the sum of human interaction within our civilization rests on a rhetorical foundation. All our communication activities can be described in rhetorical terms, whether we're educating children, reasoning or arguing with a spouse or a friend, making private or work-related decisions, listening to the words of a priest, haggling with a dealer or opposing a fellow scientist.

Some rhetorical elements can be seen as cultural conventions. Others are probably rooted in how our brain works. Take the number three, for example—three arguments, three wishes, three methods of persuasion (see below). The power of this constant application of the number three can be explained in two ways: partly as a social convention, and partly as a reflection of how the central nervous system's hardware handles information.

Some examples of rhetorical concepts in use:

1. A typical scientific journal article in the IMRaD[3] format (see Sect. 11.2) is closely related to *Dispositio* (Table 4.1)—the structure widely used for the organization of speeches in rhetoric (see Sect. 4.4).

Table 4.1 The relationship between *Dispositio* and the structure of the scientific journal article

Latin	English	IMRaD
Exordium	Introduction	Title, Abstract, Introduction
Narratio	Background	
Propositio	Thesis	
Probatio	Proof	Materials and Methods; Results
Refutatio	Refutation	Conclusions and Discussion
Peroratio	Conclusion	
		References and Acknowledgements

2. "The five Ws," commonly used in journalism, were first described in rhetoric—read more below.
3. The figures of speech we use in speaking and writing—for example "bittersweet" (oxymoron) or "brave as a lion" (simile)—were studied, developed and used within the sub-area of rhetoric called elocutio.
4. Elements of rhetoric manifest themselves in the most unexpected situations in literature and popular culture—from Shakespeare to *The Wizard of Oz*. In the latter, the three characters that Dorothy befriends on her way to Emerald City refer to the Aristotelian triad (read all about it in Sect. 4.4.1): the Scarecrow is

---

[3] Introduction, Methods, Results and Discussion.

searching for a brain (*logos*), the Tin Man for a heart (*pathos*), and the Lion for courage (= character, that is, *ethos*).

4.3 Some Glimpses into the History of Rhetoric

A fast-forward version of the history of rhetoric is not only interesting, but also useful. The development of the art and its application explains why it has managed to permeate all kinds of human activities.

The roots of rhetoric actually stretch back to prehistoric times. Examples of thinking now associated with the art can be traced back to the civilizations of Mesopotamia and Egypt. We also find elements of it in anonymous folk tales, legends, proverbs, and so on.

The Greeks

The story, as it is most often told, starts in Ancient Greece in the fifth century BC with the so-called sophists (from *sophos*, 'a wise man'), who were a class of teachers, philosophers and speakers. An educated and trained public speaker was called a *rhetor* (ῥήτωρ). Their influence reached all parts and functions of Greek society—from trade and law to education and politics. From these early days, the exploration of the most effective ways to reach an audience was intermingled with discussions about ethics, honesty, and the character of the speaker. The giant of this Greek school of rhetoric was Aristotle (see fact box). Other important names include Gorgias, Isocrates, and Demosthenes.

> **Aristotle—the rockstar of rhetoric**
> Aristotle—philosopher and polymath—is the superstar of rhetoric. He first saw the light of day around 384 BC in the ancient city of Stagira on the peninsula of Chalkidice, situated in today's province of Central Macedonia in Greece. His parents gave him the name Ἀριστοτέλης in ancient Greek, which transliterates to *Ar'isto + télēs*, 'the best purpose.' In English—which has a tendency to manhandle foreign names and words—the last part was for some reason truncated, forming the name *Aristotle*.
>
> During his brief life of 38 years, Aristotle became one of the most important intellectuals in world history—up there with geniuses like Galileo, Skłodowska-Curie and Einstein. From the perspective of what humankind knew then, he wrote not only about science (physics, biology, zoology) and philosophy (metaphysics, logic, ethics, aesthetics) but also about art (poetry, theater, music), psychology, language, economics, politics and government. A list from antiquity states that his works comprise 400 books, corresponding to half a million lines of writing.

4.3 Some Glimpses into the History of Rhetoric

> When it comes to Aristotle's work on communication, one book stands out: *The Art of Rhetoric* (sometimes called *On Rhetoric*). The original name of this book, which was compiled from writing between circa 367 and 322 BC, was Ῥητορική. Often, the Latin name is used: *Ars Rhetorica*. Reviewing its rich content here would lead too far outside the scope of this section. Its English translation reads very well and doesn't seem too alien to the modern reader. The so-called Aristotelian triad—ethos, pathos, logos—is introduced very early in the book. We recommend every person interested in communication to add it to their reading list!

The Romans
The Romans, forming a cultural branch that sprouted from Greek civilization, developed the art of rhetoric even further and produced some of the most important writings on the subject. The Latin term for a public speaker was *orator*. Just like in Ancient Greece, rhetoric was a key element in law, politics, business and war. The texts of Cicero (106–43 BC) and Quintilian (AD 35–96), for example, are still used today for analysis, research, and applied communication.

Medieval Times and The Renaissance
During the Middle Ages and onward, rhetoric was a part of the European university system and was taught within the framework of the so-called *trivium*, together with grammar and logic. Over the ages, generations of noblemen, lawyers, clergymen, and aspiring politicians learnt rhetoric and internalized it in their professional culture.

It is curious to note that one of the most famous orations of antiquity was written by Shakespeare. The speech by Marc Antony in the play Julius Caesar ("Friends, Romans, countrymen, lend me your ears …") has become so famous that it is sometimes mistaken for a historical text. It is a true masterpiece of its kind, and is often demonstrated and analyzed by teachers of rhetoric.

1800s
As the centuries passed, the theory and application of rhetoric became more schematic, formalized and stereotypical—something that created a backlash during the Romantic era when authenticity and spontaneity were lauded.

However, a refreshing example of how skilled orators could return to the origins of the art and create something brilliant is the Gettysburg Address of 1863. It was delivered by president Abraham Lincoln during the inauguration of a Civil War battlefield cemetery for soldiers from both sides of the conflict. The audience first got to hear Edward Everett, the foremost orator of the United States at that time. He talked for two hours; today, nobody remembers a single word of what he said. President Lincoln spoke for two minutes; his 271-word address is still considered one of the historical high points of political rhetoric. The speech seemed almost improvised in its wording and structure, yet it became legendary in its ingenious simplicity:

"Fourscore and seven years ago our fathers brought forth, on this continent, a new nation ..."

War rhetoric

Interestingly enough, the Second World War gave rise to a number of skilled speakers, all of whom influenced the tide of history in their respective countries. Hitler, Goebbels, Churchill, Roosevelt and De Gaulle all knew how to reach the hearts (and in some cases even the brains!) of the people they wanted to follow them into the fire. That is, with some assistance from professional speech writers, of course.

> **Famous speeches of WWII**
>
> *"... we shall fight on the beaches, we shall fight on the landing grounds, we shall fight in the fields and in the streets ..."* (The British Prime Minister Winston Churchill to the House of Commons of the UK Parliament on 4 June, 1940—after the Dunkirk evacuation.)
>
> *"Yesterday, December 7, 1941, a date which will live in infamy, [USA] was suddenly and deliberately attacked ..."* (United States' President Franklin D. Roosevelt to a Joint Session of the U.S. Congress on December 8, 1941—after the attack on Pearl Harbor.)
>
> *"Wollt ihr ihn—wenn nötig—totaler und radikaler, als wir ihn uns heute überhaupt erst vorstellen können?"* "Do you want it [the war] to be—if necessary—more total and radical, as we can imagine it to be today?" (Reichspropagandaminister Joseph Goebbels proclaims *The Total War* in Berliner Sportpalast, 18 February 1943—after the German disaster at Stalingrad.)

Modern Times

After the war and into the new millennium, well-written, engaging speeches have continued to stir attention, change the winds of opinion and—eventually—change the world. Some notable speeches written in the tradition of the orators of Antiquity, include John F. Kennedy's inauguration speech in 1961 (*"Ask not what your country can do for you—ask what you can do for your country"*), Martin Luther King's famous oration in Washington in 1963 (*"I have a dream ..."*), and Ronald Reagan's address in Berlin in 1987 (*"Mr. Gorbachev, tear down this wall!"*). More recently, Greta Thunberg garnered much attention for her speech at the UN Climate Action in 2019 (*"Yet you all come to us young people for hope. How dare you!"*).

4.4 Putting Rhetoric to Use

Rhetoric is like the rules of football or chess, the theory of music, or the genetic code: from a simple framework and a set of rules, an immense richness of applications and situations springs forth. Whether you've only read a 180-page handbook on rhetoric, or hold a professor's title in the subject, you can spend your entire life with the most basic tools of rhetoric, observing and analyzing what unfolds on the stage of human life.

But let's keep it straightforward. When you put rhetoric to use, it basically helps you with four things:

- **Observing and identifying** the discourse[4] and the players.
- **Analyzing and revealing** messages, intentions, and roles.
- **Planning, preparing, and designing** your own messages and strategies.
- **Executing** your communication activities in the most effective way.

In this section, we provide you with a set of rhetorical tools and a brief lesson on how to use them. You will then see these tools in action in the practical parts II and III of the book.

4.4.1 The Aristotelian Triad: Ethos, Pathos Logos

The Aristotelian triad—*ethos, pathos, logos*—is, in our eyes, the most powerful model for effective communication. Its three elements are sometimes referred to as the *modes of persuasion* or the *rhetorical appeals*, and, according to Aristotle, they constitute the three rhetorical means of proof that are available to convince another person.

While the above descriptions may sound very sophisticated, the actual use is plain but powerful: the triad helps us change the world through three elements: trust, emotion, and know-how. Here's how it works:

- **Ethos**[5] represents[6] the authority and credibility of the speaker, based on factors like merit, character, reputation, affiliations, and background. You may have

---

[4] Discourse is a fancy word that can mean different things in different contexts. The general meaning is "written or spoken communication," but in social sciences it often refers to the accumulated communication in a certain population on a certain topic, not only including what is being said, but also the values and attitudes expressed by the people involved.

[5] The words ethos and ethics are related: ethos (ἦθος) in its meaning 'custom, habit', forms the root of ethikos (ἠθικός), meaning 'moral, showing moral character'. However, you definitely don't have to be ethical to have a strong ethos; it all depends on the values of your followers. This has been proven again and again through the course of history.

[6] Actually, ethos is an *appeal* to the authority or credibility of the presenter. But for convenience, we're making things a bit simpler.

heard that a reliable and trustworthy person or organization "has a strong ethos." This concept corresponds to what is called a "strong trademark" in marketing.

Another aspect of ethos is identification and affinity, as illustrated by the opening line of Marc Anthony's speech mentioned above: "Friends, Romans, countrymen, lend me your ears." Creating a sense of belonging, indicating that we understand and share the needs and desires of our audience, also strengthens our ethos.

> **Science: a social system striving for ethos?**
> Science is very much about building ethos. Behind the talk of altruism, curiosity, and intellectual challenges, a large portion of the scientific community are actually striving for glory, laurels, and becoming the alpha of the group. Just look at the reward system, which is clearly built upon this! What are impact factors, university titles and research awards but vessels of pure, shining ethos? Win the Nobel Prize, and millions of people will trust you, no matter what you claim. For some individuals, it will be tempting to use the ethos from one field to build trust in another field where they might be complete amateurs.

A very cool—but also problematic—characteristic of ethos is that it is contextual and dynamic. Who is most trustworthy and in charge depends on the situation and may change quickly. For example, if the sound fails at a prestigious conference, the most respected and trustworthy person in the room can suddenly be the sound technician.

An important factor in ethos-building is coherence and consistency. One of the most despised roles a person can play is that of someone who says one thing and acts in another way: a hypocrite or an impostor. Even worse is the turncoat in times of conflict.

For the sake of ethos, a communicator should always find the right balance between appearing self-confident and acting superior. Arrogance can easily alienate people who initially agreed with you regarding the facts. This is also why you should always treat your opponent with respect, even if you find their claims preposterous. Always be polite; the moment you lose your temper and start to get snooty, your attempt to reach them will go down in flames. And the audience will certainly not be impressed by your lack of self-control.

- **Pathos** represents the feelings aroused, both by the speakers and by the audience. Skilled orators know what buttons to press to stir up a reaction, and they may also expose their own emotions—simulated or not—to boost the sentiments of the audience. Whether the feelings in question are positive (like pride or attraction) or negative (like anger or vindictiveness) is not the point. What's important is the call to action they are associated with. Or to put it differently: by being the one who controls your emotions, I can control your motivation. And by controlling your motivation, I can control your attitudes, opinions, and behavior. No matter if I wave the carrot or the stick—the result is that I make you go my way!

- **Logos** represents facts, logic, and arguments—something that aligns very well with the mindset of most scientists. In a rational society, this should really be the only legitimate means of proof.

> **Reflection IV.I**
> When on the bus or the underground with nothing to do, study the advertisements around you. How are they using the Aristotelian triad to convince you that their service or product is worth buying?
>
> *Hint: most probably you can find all three of them in each advertisement!*

4.4.2 The Canons of Rhetoric

The Romans divided rhetoric into five so-called canons—key principles, we could say in modern English. Together, these canons form a program of action for speakers, comprising invention, arrangement, style, memory, and delivery.

- *Inventio* (invention) is about collecting material and coming up with ideas for the speech.
- *Dispositio* (arrangement) involves structuring and organizing the different elements of the speech so that it is easy to follow and promotes understanding and retention. Like a piece of music, it should also have dynamics and ideally represent a composition that alternates between calmer and more intense sections—in a purposeful way!
- *Elocutio* (style) is about expressing oneself in a convincing and intriguing fashion (compare this with the English word "eloquent").
- *Memoria* (memory) is about committing your speech to memory. In contrast to the other canons, this is slightly outdated as speakers rarely learn their speeches verbatim and by heart; most people nowadays prefer to speak freely, referring to their slides or their speaker's notes. Alternatively, they use a teleprompter.
- *Actio & pronuntiatio* (delivery) involve planning the actual performance. During Antiquity, this was focused on movements (*actio*) and voice (*pronuntiatio*). To modern speakers, timing and interaction with the audience are just as important.[7]

---

[7] Posterity has sometimes tried to augment the five-part model, for example by adding a step at the beginning or a step at the end. *Intellectio* (before *inventio*) is then about analyzing the task,

Now, imagine these steps being used for preparing a presentation and you will realize that the canons of rhetoric can also be applied in a modern setting. In fact, when we discuss the planning of a presentation in Sect. 7.2, you will see that the model is alive and kicking, even in a handbook from Springer Nature in the age of social media and AI!

4.4.3 The Five Ws

The Five Ws is a group of fact-collecting servants which you can deploy into the information universe when preparing a communication activity, writing a text, solving a problem, or simply gathering key information.

Who?, What?, When?, Where?, Why?

The Five Ws have also become a favorite tool among journalists, who use them as a formula for quickly writing news items (see Sect. 8.5.2 for an overview, and Sects. 11.5 and 11.6 for practical uses). To provide more context, they sometimes add the following members to the group:

How?, With what consequences?

Historically, it was long believed that the Five Ws were first described by Hermagoras of Temnos (first century BC), who used "seven circumstances" to dissect a topic for analysis (who, what, when, where, why, in what way, by what means). However, recent research has clarified that our old friend Aristotle had written about this as well.

Through history, the Five Ws became a part of the *inventio* canon of rhetoric. A medieval handbook from England on the topic—*The Arte of Rhetorique*[8]—states:

Who, what, and where, by what helpe, and by whose,

Why, how and when, doe many things disclose.

> **Using the five Ws for science writing in outreach**
> *By James Beggs. Ph.D. (crastina.se, 2014)*
> Science writing, in some respects, is no different to other types of journalistic writing. It revolves around the Who?, What?, Why?, When?, Where? and (w)How?

---

defining the target audience and objectives, and so on. *Emendatio* (after *actio* & *pronuntiatio*) is evaluation—did the speech go as planned, and did it have the desired effect?

[8] Mair (2017).

> **Who?** It is helpful to contextualize any story. We want to know which genius is responsible for making the breakthrough; either so we can look out for their name in the future, or find out a bit more about them.
> **What**...did they discover?
> **Why**...is the discovery important?
> **When**...? We like to know when the discovery happened so we can form some idea of how long it might be until the breakthrough makes an impact—and how up-to-date the information is.
> **Where**... is this amazing research taking place? In the back of a shed or at Harvard? (Most people would usually have more faith in the latter.)
> **How**... did Who? make the What?
> My English teacher (Who?) always tried to drum (How?) the Five Ws (What?) into me at secondary school (When?) in Petersfield, England (Where?) and they have served me well ever since ...
> ... and here's Why!

4.4.4 Other Important Principles of Rhetoric

The rule of three

The number three is very common in human cultures—in catchphrases, proverbs and quotes, in jokes and fairy tales, in music and art, and so on. In religion, mysticism, and the new age, it appears everywhere. A Latin phrase states that *omne trium perfectum*[9]—'everything that comes in threes is perfect'. And in rhetoric, it is—of course!—all over the place.

> **Example of threes**
> **Catchy expressions**
>
> - Veni, vidi, vici
> - Snap, crackle, pop
> - Just do it

---

[9] Here you, attentive reader, quickly noted the meta-dimension of this three-word expression, didn't you?

> **Storytelling characters**
>
> - Three Billy Goats Gruff
> - Goneril, Regan, and Cordelia
> - Chico, Harpo, and Groucho
>
> **Existential concepts**
>
> - The Father, the Son, and the Holy Spirit
> - A legislative power, an executive power and a judiciary power[10]
> - Proton, neutron, electron

From a cognitive point of view, there is nothing really mystical about the number three. Our brain likes to find patterns and to organize things, and three is the lowest number where a real pattern starts to emerge.

You can use the number three in a lot of different ways when you communicate:

- You can plan your communication activities so that they finish with three strong main messages.
- You can divide a presentation into three parts: the start, the middle, and the ending.
- You can organize your science poster into three columns: one with background, hypothesis and experimental methods; one with the results; and one with discussion and future directions.
- Then there was something about the Aristotelian triad ...

Modesty and virtue

Scholars of rhetoric often point out that credibility is closely connected to personal character. For anyone engaging in social media and political journalism, it is apparent that audiences often focus more on who's talking than on scrutinizing the facts and the argumentation. It is a human weakness most of us share: we want the person we like (or are impressed by in some way) to be right.

According to Cato the Elder, a great communicator should combine two things: flawless character and training. *Orator est, Marce fili, vir bonus dicendi peritus*—"a speaker, my son Marcus, is a good man, trained in the art of oratory."

The *vir bonus* concept was of central significance to the Romans: a good man should strive to be, well, a good man! *Vir*, which means "man (of the free class)," has actually given rise to the English word *virtue*.

Replacing "man" with "person," the *vir bonus* idea can still guide us today. It teaches us that showing humbleness, kindness, and generosity is nearly always a good

---

[10] As we are discussing communication, it may be worth noting that sometimes the press and other media are described as the fourth pillar of democracy.

strategy for effective communication.[11] Acts like forgiving your enemies, giving voice to the disadvantaged, and making sacrifices for our common future pave the way for your other messages.

References

Drout, M. D. C. (2006). *A way with words: Writing, rhetoric, and the art of persuasion*. Recorded Books.
Leith, S. (2011). *You talkin' to me?: Rhetoric from Aristotle to Obama*. Profile Books.
Mair, G. H. (2017). *Wilson's Arte of Rhetorique, 1560* (classic reprint). Forgotten Books.

---

[11] The situation may of course be different among audiences who feel that they are under threat from an enemy outside their group. As history has proven, they may start listening to the cold-hearted, ruthless, and bloodthirsty instead. But let's not dwell on hate-speech and aggressive agitation.

Some Notes on the Psychology of Human Communication

How to Better Understand Yourself and Your Target Group

> **What you will learn from this chapter**
> Knowing a bit about how people think, and what cognitive pitfalls they might fall into, can be of great help to a communicating scientist.
> Through this chapter, you will
>
> - become more aware of yourself as a communicator;
> - learn about various kinds of cognitive bias, for example the Dunning-Kruger Effect;
> - pick up tips on how an audience's attention may be directed and retained.

5.1 How Does Psychology Link to Communication?

To succeed, every person who sets out to communicate needs an understanding of human psychology.

Firstly, we need to grasp the fundamental stuff—how information about the outer world is perceived, conveyed, and processed by the senses and the brain; this is often referred to as *perception psychology* and *cognition psychology*. On a very basic level, it's about the brain noting that, for example, something big, black and hairy is crawling across the tiles of the bathroom floor. On a more complex level, it's about the fascinating phenomenon that people watching exactly the same lecture or event will remember and describe it differently. These issues have been a part of psychology since the field emerged in the nineteenth century. Additionally, naturalists and philosophers have thought and written about it since antiquity.

Secondly, we need to understand what was earlier referred to as "human nature"—our intraspecies behavior, our beliefs about the world and our priorities when we act and speak. Inside every human, a struggle is going on. On one hand, we are molded and affected by our upbringing as well as our current cultural context. On the other hand, we are carrying evolutionary baggage in our minds—from the most basic awareness and observation skills to the most complex behaviors and interactions with our own species. The result is that we keep switching between rational and irrational behavior in the most haphazard and annoying ways.[1]

The complex dimensions of human behavior have been studied for a long time in folk tales and mythology, as well as in ancient rhetoric, literature, and art. Professional areas like sales and law have also been interested in the issue. Modern psychology had a somewhat slow start, but joined the party for real after WWII (see fact box).

> **When psychologists started to explore communication**
> After World War II, the Golden Age of Television introduced 24-h news cycles, including advertisements. The subsequent digital age catapulted the world into an information-heavy era that world history had never seen before. In this new era, communication became much more than just putting words, sounds and images together; there was a growing need to understand and engage with different audiences.
>
> This led to the emergence of a new and exciting field as psychologists wanted to understand the 'intuition' behind pre-war rhetoric. By the mid-1950s, the focus on the psychology of communication burgeoned, giving rise to concepts that are well-known today, such as "persuasibility," "personality factor," and "selective perception."
>
> It was also during this time that scientists began to realize the importance of the psychology behind communication. In many cases, they learned to apply these concepts to their own communication efforts, and improve the dissemination of their research. Nevertheless, there is still much work to be done to enhance the communication culture within the scientific community.

In this chapter, we will present some useful psychology concepts and principles that are applicable to communication. Please note that this is just an introduction to a rapidly expanding field of study; covering the entire area of established cognitive and behavioral research is beyond the scope of a book like this.

Understanding these fundamentals will help you reflect on yourself as a communicator, and seamlessly convey knowledge, attitudes, and emotions through all kinds of media—from texts and posters to presentations and videos. Additionally, these principles form an interesting bridge between traditional craftsmanship and modern research.

---

[1] This, of course, includes you, dear reader, as well as us, the writers of this book.

5.2 Reflecting on Personality

The word "personality" seems ubiquitous in communication (think: TV/radio/social media personalities, personality "shining through" someone's talk or writing, a spokesperson being the personality of a brand, and so on). Personality is indeed a useful tool for communicating your science. But what does the word mean?

For those struggling to tap into what "expressing their personality" looks like, a good definition can help out.

The Oxford Dictionary defines "personality" in two ways:

1. the various aspects of a person's character that combine to make them different from other people;
2. the qualities of a person's character that make them interesting and attractive.

Let's take this a step further and see what psychologists have to say on the topic. The American Psychological Association defines "personality" as "the enduring configuration of characteristics and behavior that comprises an individual's unique adjustment to life, including major traits, interests, drives, values, self-concept, abilities, and emotional patterns."

The fact of the matter is, everyone has a personality, and likely, you know yours intimately. But the crowd at your presentation doesn't. Nor does the visitor who stopped by your poster. What is it that they see?

To understand this better, let's head to a branch of cognitive science called "Theory of Mind." This branch investigates how humans attribute mental states to others, and how we use these attributions to predict behavior. The attributions are based on—you guessed it—people's personalities, what they say, how they act, and what can be inferred about their intentions. It's an essential social-cognitive skill. By developing a sense of what people are thinking, we can form our response in return.

What does that mean for us? Well, let's start of what it doesn't mean.

This does NOT mean you must

- change your personality
- resort to manipulation
- be pushy
- use a complicated formula
- have the power to read people's minds.

The list above frees us up from striving for an idea of perfection. Trying to be perfect—or rather "perfect," as there isn't a single Platonic ideal that we are all obliged to match—places an insurmountable level of pressure that may push us to be unnatural or fake. We end up trying too hard.

Instead, we can trust that our authenticity will resonate with the audience. When we show genuine excitement, they will sense that our project is something worth getting excited about. After dedicating so much time to honing our craft and perfecting our communication, we can have confidence in our ability to deliver effectively—and the audience will feel that confidence, too.

> **What does it mean to "be yourself"?**
> It's not easy to define what "being oneself" actually means. As a matter of fact, human philosophers have tried to find the answer to the elusive question "Who am I?" for thousands of years.
>
> To make the question less metaphysical, we can agree upon the fact that we are thoroughly socialized in a way that new "selves" pop up in different contexts and settings. During the course of a day, we play a number of different roles: the spouse, the parent, the teacher, the colleague, the patient, the customer, the child, the expert, the beginner… What's more: most of us keep fashioning a "self" that's not really us in many situations. In order to please authority, or to avoid embarrassment and social rejection, we may bend our own life rules and values.
>
> One starting point for reflection is the concept of "congruence," as defined by psychologist Carl Rogers. It refers to the alignment between a person's self-image and their true feelings and desires. It suggests that well-being and personal growth are fostered when an individual's ideal self (who they want to be) and their real self (who they actually are) are closely aligned.
>
> The authors of this book suggest the following:
>
> - You're being yourself when you feel that you don't have to restrain your personality severely to fit in.
> - You're being yourself as long as those who are dear and close to you recognize your behavior.
> - You're being yourself when you can watch yourself on a video recording without thinking "why did I act like this?" (Most of us will feel more or less embarrassed when we watch ourselves; this is not the point here.)

5.2.1 The Importance of Trust

We talk about personality—or your *ethos* (see Sect. 4.4.1 on the Aristotelian triad)—because it is one of the fundamental components of building trust.

Every audience you meet will have an inherent radar that monitors inconsistencies between what you claim and who you appear to be. If the discrepancies are minor and trivial, they may just seem somewhat confusing. However, if there is a significant mismatch, suspicion will spread through your audience, potentially leading them to turn against you. The growing distrust will hinder effective communication.

Therefore, it is vital that you take a look at yourself and ensure that your communication style and the content you deliver are coherent. Trust, or perceived credibility, often serves as the deciding factor for people to accept or reject your message. This brings us back to *ethos*—the authority a speaker gains through merit and reputation. Beyond the facts, beyond the science, beyond the issues, we must emphasize reliability and trustworthiness.

Trust and ethos have become particularly important with the rise of social media. While these platforms have great potential for engagement and democratization, they lack quality control. This raises the question of trust, especially as social media becomes one of the most, if not the most, significant sources of information. The role of scientists and science communicators as trusted gatekeepers, and their engagement on social media, has become more important than ever. Ethos-building is imperative to maintain a reliable connection between the layperson and science.

5.3 What Gets in the Way of Communication

Having discussed you, the sender of the message, we can now move on to analyze the psychological mechanisms that may hinder the receivers' understanding of your message, intentions and personality—thus making the task of building trust just that bit more difficult. While there is no point fighting these phenomena, since they are part of what makes us human, it is good to be aware of them and prepare accordingly.

Forewarned is forearmed.

5.3.1 Spontaneous Trait Inference

What is it? It's the automatic formation of impressions about people based on initial observations.

For instance? Seeing someone hold the door open for others might lead you to infer that they are a considerate and kind person. A person who slouches could be perceived as shy and lacking confidence. And in film scripting, the term "save the cat" refers to the protagonist performing an admirable action at the beginning of the story, in order to establish them as a likable person and get the audience on their side.

Why is the concept important for a communicating scientist? Hasty conclusions, drawn solely from observing a few initial behaviors, may lead to forming an inaccurate mental image. This erroneous representation can be very difficult to unpick—and may later lead to misunderstanding or even conflict. After all, you can only make a first impression once!

Spontaneous trait inference occurs when a subconscious judgment is made about another person's character traits based on observed behavior.

Let's test this for ourselves. In a 1984 paper identifying evidence of spontaneous judgment, participants were asked to read this sentence:

"The secretary solved the mystery halfway through the book."

What trait did you spontaneously infer? In the study, the majority of test subjects spontaneously inferred the trait "clever."[2]

If the initial, intuitive judgment is a negative one, it can be particularly difficult to change. Take Mr. Darcy from *Pride and Prejudice* as an example—Elizabeth Bennet initially judges him to be proud and disdainful, only later discovering his true generosity and kindness. In fact, countless book and movie plots have been based around erroneous spontaneous trait inferences.

Today, quick inferences are made casually on daily, hourly, or even minute-by-minute basis due to social media, where people spontaneously infer traits, goals, and values from very short interactions. As science communication increasingly relies on social media, or indeed is carried out entirely via social media, it's essential to understand that how you portray yourself is also how the audience will perceive your science. Studies[3,4] have shown that the more competent and moral (think: *genuine*) the communicator appears, the higher the audience's interest and engagement. Specifically in video communication, scientists who appear more "interesting-looking" and those who look like "good scientists" tend to have higher interactions.

It should be emphasized, once again, that in reality, there is no such thing as an "interesting-looking" scientist or a "competent-looking" scientist. There isn't a single "look" you should strive for. Of course, you should be mindful of your outer appearance, just as you would for an interview or when teaching children. But most importantly, it is your personality, excitement, professionalism, and motivation that will come across to people.

5.3.2 Confirmation Bias

> **What is it?** People pay more attention to information they already believe in, and actively ignore the ideas that contradict their beliefs.
> **Could you give two examples?** Selectively searching only for studies that support a hypothesis ("cherry-picking"); only recalling the positive actions of politicians from your preferred party while dismissing or forgetting their misdoings.

---

[2] Winter and Uleman (1984).
[3] *For instance*, Jarreau et al. (2019).
[4] *Or another study:* Reif et al. (2020).

> **Why is the concept important for a communicating scientist?** On some level, we are all biased. Being aware of that bias helps us communicate science with maximum objectivity and integrity. Additionally, understanding how you can comprehend and show respect for the existing beliefs of your audience will help you build rapport—and also convey facts that challenge these beliefs, if needed.

Despite their best efforts, scientists and communicators often face what is called *confirmation bias* from their audience. This means that individuals notice and focus on the ideas that reinforce what they already believe. In the process, they become reluctant to re-evaluate their beliefs or assess new or contrasting ideas with an open mind. This may even result in rationalizing concepts that are false (see box below).

The concept of confirmation bias is useful for understanding human belief systems and why our attitudes persist.[5] It can be seen as a generalizing mechanism that helps us process and organize the enormous amount of information we handle daily. Additionally, it helps us to lay a cognitive foundation on which we build our self-esteem. Just imagine the opposite: constantly challenging everything we believe in and never daring to hold anything for certain.

Due to the way our brain works, we're all susceptible to *confirmation bias*—some a little less, some a little more. Scientists are definitely not exempt, even if the scientific method is in itself designed to challenge and re-evaluate. In the process of wanting to convince the audience of the integrity and value of our science, we may succumb to overvaluing the evidence that confirms our ideas, or skillfully omitting relevant points that we feel will take away from the overall impact of our work.

> **Science people who suffer from cognitive bias...**
> ... stubbornly sticking to their opinion despite the evidence against their case piling higher and higher.
>
> - Giovanni Schiaparelli, who pointed out canals on Mars.
> - René Blondlot, a French scientist who built a career on the non-existent N-rays.
> - The international researchers who jumped on the polywater bandwagon, launched by Soviet scientists.
> - Jacques Benveniste and the idea that water has memory.
> - Physicians over two millennia who used the writings of Galen and other ancient writers as the foundation for medical practice.

---

[5] Nickerson (1998).

To avoid subconsciously 'editing' yourself, stay true to your initial goal. The idea is to communicate effectively to establish understanding in a transparent way. The best questions to ask yourself are: "Is my informative approach positive and valid?" and "Would I find my own presentation objective and unbiased if I were to peer review it?"

5.3.3 Cognitive Dissonance

> **What is it?** A mental discomfort felt by someone who faces contradictory information or feels forced to embrace conflicting beliefs, attitudes, or values
>
> **For instance?** Someone who cares about the environment and knows that car emissions are harmful to the planet might still use a car because it is more affordable and convenient. This leads to a conflict of values (environmentalism) and actions (driving), resulting in mental discomfort
>
> **Why is the concept important for a communicating scientist?** Your research might cause people cognitive dissonance—addressing it will lead to stronger and more effective communication.

As researchers, we'd like to focus on the positive aspects of science communication, where, ideally, our audience understands and is convinced by the information we present. In reality, this is not always the case. Depending on the topic, result or angle that you are trying to communicate, you may encounter pushback—whether due to lack of trust, or confirmation bias. The ensuing public reaction from the clash of beliefs, that is, cognitive dissonance, is usually one of resentment and negativity.

When cognitive dissonance is addressed, individuals feel relieved and satisfied; if not, they express dissatisfaction and frustration. People deal with cognitive dissonance in one of three ways:

(1) trivializing or forgetting ("I love animals, so I'd rather forget where my delicious juicy steak came from");
(2) changing one element of the two that are in the dissonant relationship: either the behavior ("Sunbathing may lead to skin cancer, so I will sunbathe less frequently") or the thought (denying the accuracy of the research, or finding studies that don't show a strong link);
(3) restructuring, that is adding other thoughts to minimize inconsistency ("Plastic bags are bad for the environment, but if I don't use one, I can drop and break

the items I just bought. That would be wasteful. So perhaps plastic bags are actually good at reducing waste?").[6]

As communicators, we want to focus on making a difference, so ideally we aim for the second option. Unfortunately, options two and three require much more effort than the first one—and a lot of patience, coaxing and respect on our part. Incomplete coping processes may lead to disenchantment with the scientific community.

Therefore, it is imperative to know your audience and the different perspectives they may have. Effective communication can only come from addressing the dissonances that arise in the confrontation of ideas. Look to Chap. 14 for a practical recipe to resolve cognitive dissonance and reach the disbelievers.

5.3.4 Dunning-Kruger Effect

What is it? Those who know little about a subject are more likely to overestimate their knowledge compared to those who are experts

Could you give an example? People with a superficial grasp of political/economic/biological issues may confidently express strong opinions, despite their limited knowledge

Why is the concept important for a communicating scientist? Being an expert is becoming increasingly difficult in times when everyone has access to information (of varying quality). Overcoming the Dunning-Kruger effect requires communicators to build trust and exercise patience.

Charles Darwin once said, "Ignorance more frequently begets confidence than does knowledge,"[7] and the wisdom in his words rings ever more true in an age where the world's knowledge is just a click away. A challenge that arises for science communicators is how this vast resource can negatively influence audience biases on specific subjects. The issue arises not from a failure on the part of communicators, but rather from a phenomenon called the Dunning-Kruger effect. The Dunning-Kruger effect is a cognitive bias where individuals who lack expertise in a specific area—yet are unaware of their deficiencies—overestimate their competence in that area. David Dunning and Justin Kruger, the psychologists who identified this bias, noted that even recognizing one's lack of knowledge requires a basic level of competence. What this translates to in real life is that when false information spreads to the masses, it reinforces confirmation biases among those who have not attained enough information to make the most informed decisions.

---

[6] For a review of the research on cognitive dissonance, read McGrath (2017).
[7] Darwin et al. (2004).

People exhibiting this effect often refuse to acknowledge that they are not as well-informed as they believe, posing a huge challenge for communicators.

For many scientists, the instinctive response to misinformation is to bombard the audience with facts. However, this approach is counterproductive. Instead, establishing common ground and mutual understanding, and listening to the viewpoints of the misinformed, is far more effective. People often gravitate towards incorrect information because it provides a sense of control and order in their lives. Piling on information can come across as patronizing, and your audience may feel ridiculed. It is crucial to provide the space and time for them to change their mind without losing face.

> **Reflection V.I**
> Can you recognise any of the above phenomena from personal experiences? Can you translate them into other aspects of your life?
>
> To Joanna, who spent a good few years of her life doing competitive ballroom dancing, this saying jumps out immediately:
>
> 1. Beginner dancer: knows nothing.
> 2. Intermediate dancer: knows everything; is too good to dance with beginners.
> 3. Advanced dancer: dances everything, especially with beginners.
>
> What are your experiences?

5.4 Memory, Attention, and Retention

> Memories of our lives, of our works and our deeds continue in others.
>
> —Rosa Parks.

As communicators, we face numerous obstacles, such as ignorance, boredom, or cognitive dissonance. Yet, the largest problem may be a natural one: forgetfulness. No matter how well we seem to communicate science, it often feels like we end up repeating ourselves over and over again.

Forgetting is often seen as inevitable. Psychologist Hermann Ebbinghaus was the first to conduct controlled experiments on forgetting (in the late nineteenth century), quantifying the results in what he called the "forgetting curve." His findings were inexorable, if not unexpected: information is quickly forgotten; despite all our efforts to retain information, memory degrades. It's like forgetting is the invisible villain that undermines our work.

Or is it?

In 2017, University of Toronto neurobiologists Blake Richards and Paul Frankland published a paper suggesting that memory does more than just store

5.4 Memory, Attention, and Retention

information.[8] Rather, what our brains retain is meant to assist in better decision making. According to their model, forgetting is evolutionarily beneficial, as our minds assess and discard information that doesn't aid survival.

With this in mind, the question remains: how can we preserve our work in the memories of our audience?

The answer lies in strengthening the persistence of memory.

In 2015, MIT neuroscientist Richard Cho established the concept of *synaptic plasticity*.[9] The idea is simple: neurons that fire more frequently have stronger synaptic connections, while those that rarely fire have weaker synaptic connections. The more a memory is accessed, the deeper it is encoded in the interconnected webs of neurons.

The key takeaway is that when designing our communication strategies, we must ensure our content facilitates the formation of new synaptic connections by linking to common or prior knowledge. This way, when we build on related concepts, we continue to access the memory, strengthening the recall for our audience.

Making your messages, ideas, and stories sticky

In their book *Made to Stick*, authors Chip Heath (Stanford professor) and Dan Heath (Duke University fellow) explore "why some ideas survive and others die." Building on the concept of "stickiness" by the influencer Malcolm Gladwell, they developed the acronym SUCCES—the following six characteristics can make an idea interesting and memorable:

- Simplicity
- Unexpectedness
- Concreteness
- Credibility
- Emotions
- Stories

Reflection V.II

Consider the textbox about brothers Heath and their stickiness concept. Does the content as a whole live up to the SUCCES formula? How would you improve it?

With all this in mind, we need to address one more point: before a message can be committed to memory, it first needs to capture the receivers' attention.

---

[8] Richards and Frankland (2017).
[9] Cho et al. (2015).

Yet attention is fickle—take, for instance, the attention curve: during a longer presentation, our attention is highest at the start, then takes a deep dive, and rises slightly at the end. And the attention span has, according to reports from Microsoft, decreased from 12s at the beginning of the millennium to 8s fifteen years later.[10] The audience's attention needs a bit of a leg-up—and we will now look at a couple of principles that may be of assistance.

5.4.1 Directed Attention—Imperative in All Kinds of Communication

> **What is it?** The assignment of attention to specific information or cognitive processes in a controlled manner
> **Such as?** When a conference speaker points to a detail in a graph, the audience can focus on it to understand its significance
> **Why is the concept important for a communicating scientist?** Because communication is only effective if the audience is paying attention to what you want to convey. As the saying goes: "I want to make sure that we're all on the same page."

Listen up, I have some really useful knowledge for you! Imagine a lecture hall where no one is paying attention to you as a speaker. Depressing, isn't it? Worse than that, it is a total waste of time for both speaker and audience.

However, initiating the communication activity—that is, opening up the channel between *Sender* and *Receiver* of the Shannon-Weaver model—is only the beginning. Once the activity has started, a skilled communicator continuously ensures that both sender and receiver focus on the same details: *"In Figure 1, you can see ...,"* *"Look at this gray spot in the X-ray image,"* or *"May I turn your attention to the difference between these two bars..."* ...

Directing attention can be done in a multitude of ways, depending on the medium.

- During a lecture, the speaker may use gestures to indicate what's being described (see Fig. 5.1).

---

[10] Except it's not actually that simple. These figures have been widely disputed, since they merely reflect how much time a person spends examining a website before moving on. For an interesting account of attention—and how to hold it in a lecturing context—look at Bradbury (2016).

Fig. 5.1 Skilled weather presenters use their whole body to guide the audience across the map. By turning to the map after their introduction, they signal a switch to a geography-focused approach. With their hand gestures, they direct the viewers' attention to what they are discussing at the moment. [*Image by* Mohamed Hassan from Pixabay]

- In an infographic, an arrow or a frame in a contrasting color may direct the viewer to an important detail.
- In a text, the word **Summary** in bold letters highlights the most prioritized information.[11]

A professional area where practitioners excel at capturing attention is advertising. Just take a look at a typical fast food chain poster: at the center is a delectable burger, surrounded by crunchy fries and a refreshing drink. The lighting, background color, font type, font size, and font color are all selected to sustain the viewer's attention: *"Look how delicious this burger is; you know you want to eat it."* This kind of tantalizing attention grabber harnesses the audience's focus from the very first instance, even in a world full of distractions.

---

[11] In these examples, the principle of salience is also at work, see the next Sect. 5.4.2.

When communicating science, directing the audience's attention is a core skill. We will revisit this topic below, as well as in the sections on giving presentations (Chap. 7), designing lecture slides (Sect. 12.2), and creating scientific posters (Sect. 12.3).

> **What *is* interesting, anyway?**
> *Interesting* is one of these words that are widely used without being properly defined (see Sect. 3.1.1). It is derived from the Latin prefix *inter-* 'between' and the word *esse* 'to be'. Dictionaries generally define it as 'engaging' or 'being able to get and hold the attention of someone'.
>
> We think these explanations are a bit shallow, and suggest the following instead:
>
> > *Interesting: something that either confirms, expands, challenges, or contradicts your knowledge, experience, or opinion of something that either (1) is important or relevant to you, or (2) brings joy and meaning into your personal world.*
>
> The former area of interest can guide you in determining the *needs* of your audience. Usually, the solutions to those needs are what make your science communication interesting. Businesses also use this technique to gather support and interest around their products.
>
> The need in question may not be immediate or very apparent. Take climate change, for example. Addressing climate change is essential now more than ever, yet the topic is as misunderstood or miscommunicated as they come. Focusing interest on such a topic may begin by addressing the question, "Why should we care? Why is this important?" Give the audience a *reason* to listen. It may be that your proposed answer or solution is what brings people to your door—and compels them to stay.

> **Reflection V.III**
> Have a closer look at the definition of the word *interesting* above. Compare it with some definitions you find in dictionaries and on the web—in English or in your own language. Is there anything you would like to modify, add, or remove? If you'd start from scratch, what would your definition look like?

5.4.1.1 Principle of Salience

What is it? The quality of appearing particularly noticeable or important—stimuli that stand out are more likely to capture attention
Could you give some examples? Bright colors, loud noises, distinctive shapes, unexpected actions
Why is the concept important for a communicating scientist? It's one of the most important tools in the toolbox to guide attention, demonstrate relations, and tell stories.

"Difference detector"—this epithet is given to the human brain by the cognition psychologist Stephen Kosslyn. He refers to our amazing ability to perceive things that stand out or are out of the ordinary. In his book *Clear and to the Point*,[12] Kosslyn demonstrates how this can be used in all areas of visual communication: an audience always tends to give attention to apparent differences.

In perception psychology, salience is the quality of prominence, of being particularly noticeable. Perceiving salience is the basis for communication, interaction and learning—essentially, for safely navigating the world.

We have seen that the planning of science communication starts with identifying the core messages. While the communication may also include background, reflections and other components, make sure to align your use of salience specifically with your core messages.

Reflection V.IV
In Fig. 5.2, the principle of salience is at work. Interestingly enough, it results in two alternative ways to direct the attention. Which examples shout "Look here, look at me!" and which shout "Look there, look at that!"?

---
[12] Kosslyn (2007).

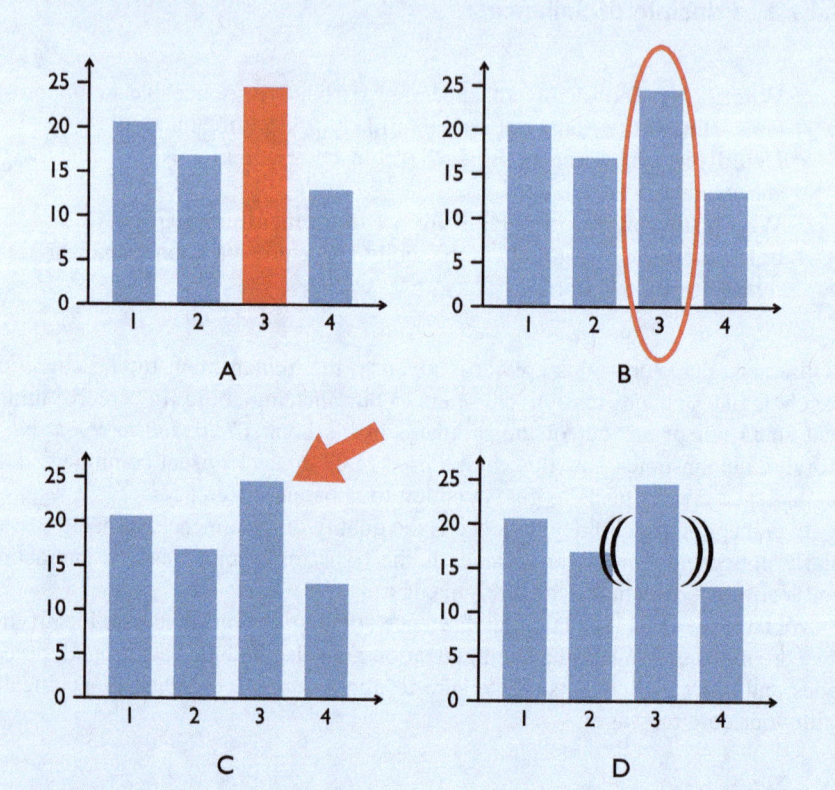

Fig. 5.2 Using the principle of salience to draw the viewer's attention to Sample 3 in the bar chart. A contrasting color (**a**), an attention-grabbing outline (**b**), or a chubby arrow (**c**) does the job, as does the wiggling of the bar (**d**)

References

Bradbury, N. (2016). Attention span during lectures: 8 s, 10 min, or more? *Advances in Physiology Education, 40*(4), 509–513. https://doi.org/10.1152/advan.00109.2016

Cho, R. W., Buhl, L. K., Volfson, D., Tran, A., Li, F., Akbergenova, Y., & Littleton, J. T. (2015). Phosphorylation of complexin by PKA regulates activity-dependent spontaneous neurotransmitter release and structural synaptic plasticity. *Neuron, 88*(4), 749–761. https://doi.org/10.1016/j.neuron.2015.10.011

Darwin, C., Moore, J. R., & Desmond, A. J. (2004). *The descent of man, and selection in relation to sex*. Penguin Books.

Jarreau, P. B., Cancellare, I. A., Carmichael, B. J., Porter, L., Toker, D., & Yammine, S. Z. (2019). Using selfies to challenge public stereotypes of scientists. *PLoS ONE, 14*(5), e0216625. https://doi.org/10.1371/journal.pone.0216625

Kosslyn, S. M. (2007). *Clear and to the point: 8 psychological principles for compelling PowerPoint presentations*. Oxford University Press.

References

McGrath, A. (2017). Dealing with dissonance: A review of cognitive dissonance reduction. *Social and Personality Psychology Compass, 11*(12), e12362. https://doi.org/10.1111/spc3.12362

Nickerson, R. S. (1998). Confirmation bias: A ubiquitous phenomenon in many guises. *Review of General Psychology, 2*(2), 175–220. https://doi.org/10.1037/1089-2680.2.2.175

Reif, A., Kneisel, T., Schaefer, M., & Taddicken, M. (2020). Why are scientific experts perceived as trustworthy? Emotional assessment within TV and YouTube videos. *Media and Communication, 8*(1). https://doi.org/10.17645/mac.v8i1.2536

Richards, B. A., & Frankland, P. W. (2017). The persistence and transience of memory. *Neuron, 94*(6), 1071–1084. https://doi.org/10.1016/j.neuron.2017.04.037

Winter, L., & Uleman, J. S. (1984). When are social judgments made? Evidence for the spontaneousness of trait inferences. *Journal of Personality and Social Psychology, 47*(2), 237–252. https://doi.org/10.1037/0022-3514.47.2.237

Part II
Learning Core Skills

"Grau, teurer Freund, ist alle Theorie und grün des Lebens goldner Baum."
('Gray, dear friend, is all theory, and green the golden tree of life.')

—Johann Wolfgang von Goethe

Welcome to Part II and the realm of everyday experience! Enough talking about the inner workings of the models and theories—it's all about practice from now on. We will focus on the most common communication activities in your professional life as a scientist, which include tasks like giving presentations, writing texts, and designing slides and other visuals. By learning, practicing, and eventually mastering these core skills, you will be better prepared to succeed as a researcher in a transforming world.

With these skills as your foundation, Part III will then become your extended toolbox, guiding you whenever you face communication activities in a range of contexts, purposes and channels.

Finding the Right Starting Point for Any Communication Activity

Frameworks for Any Task, Context, Based on Shannon-Weaver, Rhetoric, or Storytelling

6

> **What you will learn from this chapter**
> Throughout this chapter, we take principles discussed in Part I, and demonstrate how to apply them in your communication activities. Specifically, we will:
>
> - detail how the five universal guidelines can be used when prepping for a range of communication formats;
> - show how ethos, pathos and logos fit into each of the universal guidelines and how they can enhance and solidify your planning considerations;
> - use storytelling techniques to transform whatever you want to convey into a captivating and memorable story, whether it be in three minutes or an hour.

6.1 Setting Up a Plan

Whatever we set out to do, getting started is often one of the trickiest parts. When your profession involves advising others on how to perform different tasks, one of the most common questions you hear is "Where and how do I start?" Therefore, this chapter contains three sections that serve as helpful starting points for any communication activity:

6.2 The Five Universal Guidelines
   An all-round, effective, and professional way of approaching communication.
6.3 The rhetorical elements: ethos, pathos, logos.
   The time-tested Aristotelian triad at your service.
6.4 The basics of storytelling
   A communication method that has co-evolved with *Homo sapiens*.

> **Making your ideas people-friendly**
> Three tips from Magdalena Bibik, idea strategist and founder of bibik Co. Magdalena specializes in helping people turn ideas into successful products/services and in creating systems for idea generation.
>
> 1. People will use what you created in a way that suits them, which might not be the way you thought or predicted. Embrace it—and learn from it.
> 2. Understand that every new idea represents a risk. You have spent tons of hours around your idea, but someone else sees it for the first time. Be patient.
> 3. Simplify your communication. And then simplify it again. Focus on the result your idea offers, rather than features and technical specifications. Remember to communicate the right result for the right counterpart (user, buyer, investor).

6.2 Five Universal Guidelines and How to Use Them

In Sect. 3.2.1, we used the Shannon-Weaver model as a starting point for a practically-oriented analysis. The result was a set of *Five Universal Guidelines* representing an effective and professional way of approaching communication. In this section, we delve deeper into these guidelines, and demonstrate how they can be applied, pretty much universally, to a variety of communication activities.

Guideline 1: Define your goal and your main messages

Every time you are in the process of communicating, ask yourself two things. Firstly, why are you doing this, and, secondly, what are your main messages?

(A) **Why are you doing this?**

What's the purpose of your efforts—what do you want to achieve, and what result do you want to see? Or, more bluntly: why should the audience care about your text, presentation, poster, or illustration? Here are some reasons that continuously appear in the life of a scientist:

- **Transfer information**

  Do you want the information stored in your brain to find its way to other people's brains—or at least convey the basics on how to approach a subject?

- **Create understanding**

  Are you a teacher? In this case, you cannot just commit to one-way lecturing. Instead, you need to have a dialogue where you ensure your audience forms insights and develops more advanced thinking about the subject you teach.

- **Motivate people or bring members of a group closer**
  Do you want people to engage in your project or follow you into the fire? Or as Shakespeare put it: *"Once more unto the breach, dear friends, once more*[1]*!"* Then the toolbox of ancient rhetoric may prove useful (see Chap. 4) Shakespeare (2020).

- **Promote an idea, convince an opponent, or secure funding**
  Again, your persuasive skills are put to the test. In this case, it may be interesting to have a look at sales techniques; there's not much difference between advocating for a concept and selling a product.

- **Influence decisions**
  In many countries, researchers from STEMM subjects are poorly represented in the political system, policy-making, and governance. To change that, science people should be more proactive about ensuring that the results of their work can be applied in practice.

- **Make a difference!**
  Whatever the purpose of your communication activities, there should always be a greater aim to change the course of affairs of Life, Universe, and Everything. The result may be seen right away—or it may take years.

(B) **What are your main messages?**
With the above in mind, imagine that you only have the time to convey three statements to the audience. What would they be? Alternatively, think "If they can only remember three things at the end of this exercise, what would I want these three things to be?" These are your main messages (also called "take-home messages," "key messages," or "the bottom line"[2]).

Defining your goals and your main messages—some examples
Setting clear main messages makes every communication activity more effective. What's more, setting them early in the process makes planning and preparation faster and simpler.

1. **Video tutorials**

    Let's start with something very practical. The competition on YouTube and TikTok has turned the platforms into a never-ending repository of examples of brief, well-planned and transparent communication. Successful tutorials in this context—be it fashion, DIY, crafts or any other

---

[1] From Henry V (1599?) by William Shakespeare, spoken by King Henry when the English attacked Harfleur in France (Act 3, Scene 1).
[2] The phrase "bottom line" originates from the field of accounting, where the final figure on a financial statement answers the main question: "profit or loss?"

subject—are built on robust and clear goals and main messages. For instance:

Goal
To show an easy way of making a stylish and functional ponytail.

Main messages
- Lift your hair at the base before tying it back.
- Use a hairband in a similar color to your hair.
- Tuck a bobby pin in the middle to achieve a fuller ponytail.

Et voilà!

2. **An open lecture about urban wildlife**
   Our second example is from the field of outreach. Many scientists recognise the value of engaging a wider public in data collection through citizen science projects. Much like video tutorials on YouTube, these also work best with a clear-cut goal and messaging.

Goal
To engage the audience in a garden biodiversity survey.

Main messages
- Population monitoring is key for biodiversity conservation.
- Gardens are an important refuge for wildlife in urban areas.
- You can easily report the species sighted in your backyard via a biodiversity monitoring app.

Guideline 2: Know your target group

As noted in Sect. 1.3, today's scientists don't just work towards one single target group, but actually encounter a whole range of different audiences, including their peers, funders, politicians, businesses, the media and the public.

Knowing *who* they are is just the start—you also need to understand what they find interesting and useful, what motivates them, and what their values are. Who are you *actually* talking to? Why should they have an interest in you? Remember: a great communicator is a great observer and listener (Fig. 6.1).

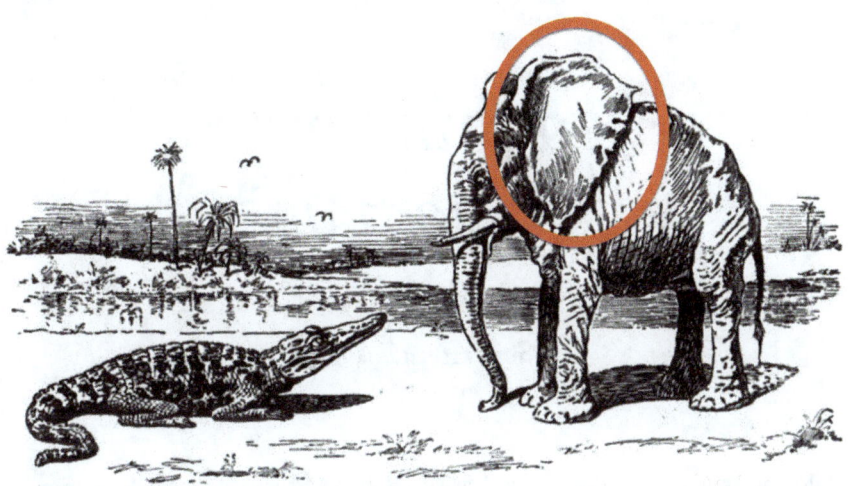

Fig. 6.1 A great communicator is a great listener! The crocodile versus the elephant metaphor is popular in management courses. The question posed to the audience is: "Which animal was born to communicate effectively?" The correct answer is the elephant; it has big ears to listen with, but a (relatively) small mouth to speak with[3]

The clearer you can visualize your audience, the more effectively you can adapt your use of channels and messages for optimal effect. Don't just think "The Media." Instead, consider: "Is it the BBC or Fox News, a local newspaper or national radio?" Similarly, what is actually meant by "The Public"? Is it your mum, your hairdresser, or your primary school teacher?

Guideline 3: Know yourself

To appear trustworthy and authentic (see Sect. 5.2.1), you should always use yourself as the starting point: your own personality, passions, interests, and values (Fig. 6.2).

---

[3] As the right answer is metaphorical rather than scientific, we have to raise a warning: presenting this illustration to a group of zoology students does not achieve the desired effect.

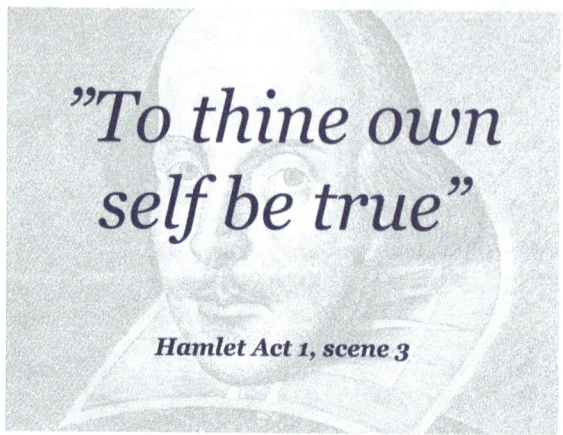

Fig. 6.2 Shakespeare holds the key to authenticity. Marcus Aurelius and Simone de Beauvoir had similar ideas, but nobody put it more succinctly than Shakespeare. The quote is from the scene where Polonius gives Laertes advice on how to behave when attending university. Polonius' speech ends thus: "This above all: to thine own self be true, and it must follow, as the night the day, Thou canst not then be false to any man."[4]

So, what is your communication style? That is, how do you encode information for best effect? And how do you act on and handle the feedback you receive?

> **"Being yourself" when communicating**
> We discussed what it means to "be yourself" in Sect. 5.2. Behaving like your "true self" in a public sphere is probably one of the hardest social skills to master; it can take years, even decades, to find the necessary self-esteem and human insight. What's more, audiences can smell posturing from a mile away. So, don't try to be somebody you are not; for instance, don't try to be funny if humor doesn't come naturally to you.
>
> Instead (Joanna usually gives this piece of advice to her students during presentation classes), be yourself, but the best version of yourself—that is, 110% yourself! Be a bit more outgoing, a bit funnier, a bit louder. Amp up your personality—just enough to be memorable, rather than annoying.

Guideline 4: Understand the limitations at hand
Every communication activity is limited by time, bandwidth, available technology, background knowledge, language skills, venue size, number of participants, and other factors (Fig. 6.3).

---

[4] Shakespeare (2011).

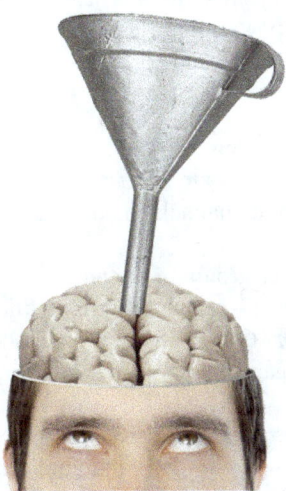

Fig. 6.3 One type of limitation that is especially important to stress for science and tech people is cognitive constraint. Although capacity varies widely between individuals, there is always an upper threshold for how much information people can absorb per unit time. For this reason, you should, for example, not show multiple graphs in one slide; instead present them one by one, in a sequence

This is something you learn more about as you gain experience. Therefore, senior communicators typically focus on limitations already at the start of the planning process, whereas juniors tend to focus on opportunities and possibilities.

- Act like a *senior* if you are looking for an effective, smooth, and safe process. The two-thirds rule (see Sect. 7.2) is a typical example of this way of thinking.
- Act like a *junior* if you want to create opportunities for something disruptive and ground-breaking (although the risk of wasting your time or failing is higher).

Guideline 5: Seek inspiration in all types of communication.
Traditionally, science has used very few channels for communication; the traditional ones are scientific journal articles, oral presentations and scientific posters. As the number of available avenues grows, science people ought to look for inspiration from all channels, modes, and genres of communication—including teaching and the arts. This includes:

- advertising, marketing and sales;
- journalism;
- PR;
- literature;
- drama, theater and comedy;
- music.

Writing fiction can enhance your skills as a scientific writer; exploring theater arts may help you deliver more powerful presentations; studying advertising techniques can improve your poster design—and so forth.

> **Expanding your communication portfolio**
> Here are some tips on how to extend your communication mastery by studying and practicing different channels, modes, and genres of communication.
>
> - Join an improve theater group, or attend a stand-up comedy course.
> - Produce popular science—either as written articles (e.g., for the not-for-profit media outlet *The Conversation*[5]), podcasts, or videos.
> - Organize a public engagement activity with primary school children.
>
> All of these—and many more—will give you exposure to different types of communication, improving your skills and confidence as a result.

6.3 Putting the Rhetorical Elements Ethos, Pathos, Logos to Use

The Aristotelian triad (Table 6.1)—ethos, pathos, logos—was described in Sect. 4.4.1. Here, we loosely superimpose it on the Five Universal Guidelines that you are already familiar with, creating some tips on how to put the two to use. As you can see, the recipe is not particularly formalized, and we admit that the approach is a bit unorthodox (rhetoric purists have been warned!). That said, we ask you to free your mind, open your eyes, and analyze the communication task by responding to a set of questions.

Table 6.1 The Aristotelian triad cheat sheet

Ethos	Trust, believability, prestige, interest, popularity, title, track record, professional role, affiliations, connections, personal trademark, organizational brand
Pathos	Positive and negative feelings, appeal, enthusiasm, panache, certainty, revulsion, fear, anxiety, hostility, threat, uncertainty, relief
Logos	Facts, logic, relevance, methodology, statistical significance, "common sense"

- If the question doesn't give a useful answer, just put it aside.
- If the question leads to an interesting observation, insight, or doubt, write it down and consider it during your planning.

---

[5] The Conversation Media Group Ltd. (2024).

6.3 Putting the Rhetorical Elements Ethos, Pathos, Logos to Use

Let's take a look.

1. **For your communication activity: define the goals and the main messages.**
   Easy—see Sect. 6.2 above.

2. **Analyze how *the audience* appears to you.**
   Star scientists, seniors, same-age peers, students, or school kids—how does the audience's level of knowledge, experience, and prestige affect you and the task at hand?

   The ethos perspective
   What kind of authorities does the audience usually trust? Do you feel intimidated, and consequently downplay (or big up!) your own importance or competence? Or is the opposite true: do you feel you have a strong position—professionally, culturally, socially—compared to the audience?

   The pathos perspective
   Does the culture of your audience allow for a discourse about feelings, or should you keep a stiff upper lip? Alternatively, is the audience so emotionally focused that it is challenging to communicate from a rational standpoint?

   The logos perspective
   Do you understand how they perceive the world, and what proof they find valid? Do they use your common language in the same way as you do?

3. **Analyze how *you and your message are perceived* by the audience.**
   Step into the shoes of the target group and reflect on their expectations, preconceived notions, and prejudices. Ask yourself "*What do I, my organization and/or my main message represent to them?*"

   The ethos perspective
   Do they already know me? Do they trust me as a peer? Do I belong to their tribe—am I one of them? Or do they consider themselves superior, or perhaps inferior? Am I expected to continuously put the proof on the table, or do they believe what I say upfront? Are they perhaps suspicious—judging me, waiting for me to make a blunder? Alternatively, do they admire me or look up to me in a way that may inhibit our interaction?

   The pathos perspective
   What kind of feelings do my subject and my messages evoke among the target group? Will they become enthusiastic, attentive, turned off, alienated, hostile, or belligerent? What type of emotional manifestations are considered inappropriate or unserious? How will the expected emotions affect my communication purpose, and how can I leverage (or mitigate) these effects?

The logos perspective
Do they have the background to understand my usual way of explaining things? Or should I find the right level by simplifying my material, or perhaps elaborating on it?

4. **Analyze how *the context* influences the communication process.**
   From a rhetorical point of view, this part is a bit unorthodox and may not always provide useful information. However, if it does, it can distort your interaction with the audience in the most unexpected ways. Here are some examples:

   The ethos perspective
   - The surroundings themselves can make people starstruck and inhibited (e.g., a top conference in the field, the campus of a prestigious university).

   The pathos perspective
   - A colleague just walked by and people moved their attention elsewhere.
   - Recent failure has jinxed the project you are discussing.

   The logos perspective
   - Current headline news affects the topic of your presentation.
   - An important article or report has been published, which many players now want to bring into the discussion.

5. **Merge the results of your analysis to find the most effective strategy.**
   Or rather: the most effective combination of strategies. It is very rare to use only one mode of persuasion.

   Convincing them with ethos
   You must always create some kind of ethos-based foundation for every communication activity. Without ethos, there is no communication. Some aspects include:
   - Coherence with what they already know and think.
   - Shared values and role models.
   - Common references regarding profession, education, culture, leisure activities, social platforms.
   - Professional and personal connection.
   - General trustworthiness, respect, authority.
   - Track record and expertise.

   Convincing them with pathos—appealing to the audience's emotions.
   Since the readers of this book will mostly deal with university and business audiences, a word of caution: when adding emotions to your persuasive mix, always do so in a way that appears appropriate for the audience in question. Consider:
   - Concerns about our shared future—and how to find solutions.
   - Success stories of perseverance and bravery overcoming adversities.

- Generosity, tolerance, kindness—the strong respecting and caring for the weak.
- Human suffering, poverty, disease—and how to alleviate it.
- Family ties, community solidarity, team spirit.
- Diversity, equity, inclusion.
- Anger and indignation towards a lack of reason, moral values, justice.
- Joy of finally solving a well-known scientific challenge.

> **The authors reflect on culture and emotions**
> What emotional expressions are considered appropriate during, for instance, a presentation? How personal can you be? This differs greatly between cultures. Here are some subjective reflections from the authors.
> *Olle is dyed-in-the-wool Swedish, but has an extensive international network*:
> "Swedes may seem strangely detached in social life, but are more tolerant towards the emotions—unexpected or personal—than you may think. However, it's usually more fun to have an audience from southern or eastern Europe than from western or northern Europe. Being a person 'from another place' often makes people more tolerant; this is very useful for a traveling lecturer!"
> *Joanna is born in Poland, raised internationally, and now lives in the UK*:
> "In my experience, Brits are incredibly uncomfortable when someone shows emotions. In contrast, Americans are very much ok with them. Poles are fine with emotions—even strong and passionate ones—as long as they are the emotions they expect. Accordingly, tears at a funeral are fine, but jokes in a science lecture don't always go down well. Trust me, I've tried!"

Convincing them with logos.
This is where most scientists feel most confident. Let your science shine! We employ:
- Logical arguments: if A, then B ...
- Displays of scientific data and statistics from trusted sources.
- Conclusions and compiled facts from trusted sources.
- Quotes from trusted sources.
- Photos and figures.

> **A note on anecdotal evidence**
> It is hard to say where anecdotal evidence belongs in the ethos/pathos/logos universe. Often introduced as an appeal to logos, it may not live up to the standard criteria of scientific proof. On the other hand, a single,

> well-documented observation can serve as proof of concept, demonstrating that something is worth a closer look—thus building some ethos for the speaker. Finally, it can appeal to pathos by showing a more subjective, experience-based side of the researcher.

6.4 Telling a Good Story

Recall the last time you heard a truly excellent story. Maybe you were sitting around a campfire or in the comfort of a beautiful room? Remember how your eyes were fixed on the storyteller, their intriguing voice weaving through the air, capturing your attention:

옛날 옛날에... *Once upon a time... Det var en gång ...* 很久很久以前... *Na ishte një herë... Dawno, dawno temu...*

Stories can just as well be served at the coffee machine at work or in a university café. Perhaps a coworker bombs you with the latest, juiciest office gossip: "*Hey, have you heard...?*" Or a friend plops down dramatically, sighing, "*I can't believe he/she/they...*"

What is it about stories that makes them so intriguing? Why do we enjoy them so much, and why are we continuously drawn to them?

It's not too bold to claim that since the beginning of time, mankind has told stories. Going far beyond trivial entertainment, stories have been used to explain humanity's place in life, or the mysteries of the universe. Stories absorb us, allowing us to live a mental, emotional, and visual experience that is not our own.

But telling stories also has practical functions. Stories can illustrate abstract concepts, bring facts to life to enhance understanding, and walk the listener through the mind of a scientist. They can persuade listeners, helping them overcome resistance or anxiety to new ideas. Additionally, stories direct and sustain attention. In fact, such is the power of a storyteller that it is now a centerpiece skill needed in every industry. Nick Morgan, author of *Power Cues*,[6] points out that it is the stories that create the "sticky memories"[7] which allow factual information to stay in our heads. He goes further, stating that you "won't be heard unless [you're] telling stories."

> **Reflection VI.I—The fact-to-story ratio**
> Facts are like broccoli. Broccoli is good for you, and having a bit at a time is fine—but a meal made up entirely of broccoli (see Fig. 6.4a) ends up rather dry and unpalatable. On the other hand, adding other ingredients to

---

[6] Morgan (2014).

[7] Compare the box about stickiness in Chap. 5.

broccoli—perhaps a bit of cheese, herbs, potatoes, salt, butter—makes for a much more appetizing and wholesome eating experience (see Fig. 6.4b). In your case, these extra ingredients are the story. As a communicator, you need to make sure that the fact-to-story ratio is not off-balance.

Fig. 6.4 The broccoli/fact analogy summarized in photographic form. [*Images by* Andi and Bernadette Wurzinger from Pixabay]

We scientists can greatly benefit from using stories. Not only do listeners retain plot lines better than data, but incorporating a plot creates a trajectory for us to follow, helping us avoid over-detailed, bland presentations. The benefits of telling stories confidently are truly worth the effort.

The first question: "Do we *need* to tell a story?"

Storytelling is but a tool. It's neither the best, nor the only method of instruction. Rather, it's an enhancer that can be added at a strategic moment in a communication activity (written, spoken or visual):

- at the **beginning** to gain attention
- in the **middle** to introduce a complex idea
- at the **end**, to summarize and wrap up the talk.

Review your content and identify points where a story could effectively convey your messages.

Select a story appropriate for the task at hand.
As per Sect. 6.4, determine the parameters of your goals, identify your audience, and the intended purpose of your presentation. The answers to your pre-prep will help you decide the kind of presentation you want to give or the article you want

to write. These answers are also important to consider as you peruse your selection of stories.

What do we mean by "your selection of stories"? This is your story repertoire, which may consist of:

- your own life experiences;
- stories from your family (these can go back decades or more);
- anecdotes from acquaintances or friends;
- stories you've heard, read, or watched that left a lasting impression—this includes everything from anonymous schoolyard jokes to plots in novels and films.

The stories we tell best are those from our own life experiences; these are usually the most effective. Stories of success and triumph, failure and defeat, overcoming hardship, or finding community resonate with everyone. To this end, we suggest keeping a "story database." It doesn't take too long—an hour can pass quickly as you jot down notes of your life's high and low moments. You can write them in a notebook or on your phone. Then, as you develop your talking points, match your key messages with a story from your database.

Stories are ancient memes, and we all know what a meme is: a form of online content—for example a joke, an image, or video—characterized by its humorous nature, rapid dissemination, and adaptability through remixing and evolution. In a broader sense, it represents an idea that swiftly propagates within a given cultural context. It's therefore worth highlighting that you don't have to limit yourself to *your personal* stories. If someone tells a good story and you can retell it well, use it! Spread it around!

When telling any story, consider its structure.
All stories have structure. Or, all stories ought to be told with structural integrity. If they aren't—well, you know what happens! We've all seen terrible movies that leave us more confused than when we started. Or attended presentations that made us feel like we've wasted our afternoon. Or half-slept through lectures where the daydream was more interesting and fun. If you tell a story with no structure, there will be no interest or impact—plus you could lose credibility. Your audience would most likely rebel in one way or another.

The most compelling stories follow the Aristotle's Three Act structure (this guy again! we know him from Chap. 4):

- **Act 1**: Sets up the exposition (background, main characters, setting) and leads us to the inciting incident, where the main character experiences a call to action that sets them on their journey.
- **Act 2**: Conflict, whether it be risk, danger or struggles, dominates here. The character faces accumulating internal and external issues they must resolve, moving towards the central conundrum that pushed them on the journey in the

first place. The tension is pulled, taut—finally, it snaps in the climax—setting up the final act.
- **Act 3**: Here, the story wraps up, and the main characters find their place back in their old life—or maybe they don't. Regardless, the story ends.

If we were to draw this structure—very roughly—it would resemble a slightly distorted letter W[8]—see Fig. 6.5.

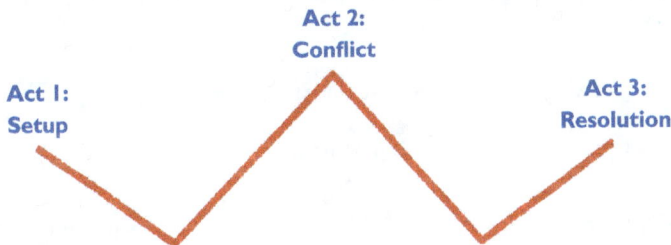

Fig. 6.5 The Three Act story structure resembles the letter W

This W-like structure not only aligns well with the Rule of Three (see Sect. 4.4.4); it also crops up in a number of varied settings. Some are very general (think of any story, article, presentation)—see Fig. 6.6.

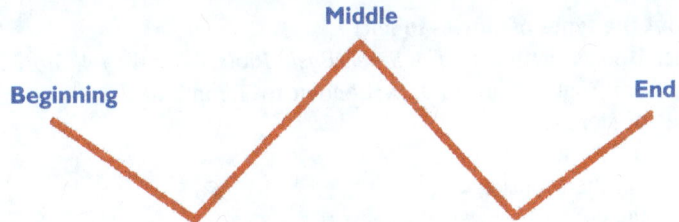

Fig. 6.6 A more general narrative structure also resembles the letter W

Some more specific (think of Steve Jobs' iconic speech introducing the iPhone in 2007[9])—see Fig. 6.7.

---

[8] In U.S. schools, this structure is referred to as "plot mountain," because of the rising action from Act 1 until the climax in Act 2 (peak of the mountain), and the falling action in Act 3.
[9] Protectstar Inc. (2013).

Fig. 6.7 A business pitch, innovation announcement or grant application can, once again, resemble the letter W

And some are very familiar to all scientists (see Fig. 6.8).

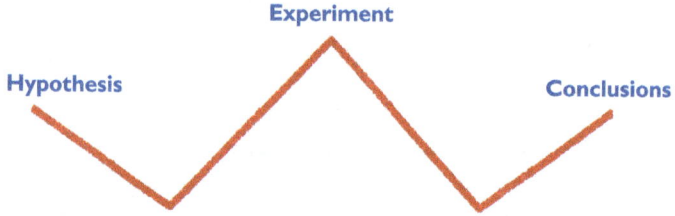

Fig. 6.8 The write-up of a study resembles—have you guessed it?—the letter W

What about the types of stories to tell?

Christopher Booker, author of *The Seven Basic Plots: Why We Tell Stories*,[10] suggests that from Shakespeare to Jack London to Tarantino, there are only about seven types of stories:

1. Overcoming the Monster
2. Rags to Riches
3. The Quest
4. Voyage and Return
5. Rebirth
6. Comedy
7. Tragedy

Of course, there are other proposals of story types, such as the Hedonometer's six story types, or Foster-Harris' three story types. But here, we will focus on four story types from Booker's list that are most relevant to science communication, primarily because they involve a definitive solving of an issue.

---

[10] Booker (2004).

1. **Overcoming the Monster (examples: James Bond movies or *Jaws*)**
   In this story type, the main character is in an idyllic setting where everything seems calm and well. Suddenly, a great evil—or struggle, calamity, you name it—interrupts the perfect day. Our hero travels far and wide, defeats the evil, and restores peace to the community.

2. **Voyage and Return (examples: *Wizard of Oz, Alice in Wonderland*)**
   The hero is suddenly placed in a new, unfamiliar setting, feeling out of place and uncomfortable. At every turn, danger—or worse, the risk of making mistakes—is rife. Still, through perseverance, the hero finds a community that helps them settle in the new environment. Often, the hero returns to the place of origin; otherwise, the new, strange land becomes home.

3. **Rags to Riches (examples: *Aladdin, Cinderella*)**
   In this story type, the hero starts in a lowly position, overworked, underpaid, and aspiring for more in life. They encounter an opportunity that allows them to acquire wealth (literal or figurative), enabling them to overcome their main adversity. Through hard work or good character, they establish themselves as a reliable person of authority.

4. **The Quest (examples: *Lord of the Rings, Harold and Kumar Go to White Castle*)**
   In this story type, the main character is not our hero. Rather, the journey is the hero. The quest is educational, pushing the main character through trials and tribulations, fostering growth. The real purpose of the quest is self-discovery, as the main character uncovers new facets of who they are.

How fun and creative can I be?

How entertaining should you be? Like Monte Carlo: the more you risk, the more you can win.

Of course, your story should be tailored to help your audience understand the point you are trying to emphasize. You don't want the story to be so overwhelming that it runs away from you. Be creative, yes, but also be strategic.

Let's say you are researching spinal muscular atrophy (SMA), a rare neuromuscular disease that, if untreated, is the most common genetic cause of infant death. When discussing it, you could simply present the facts: the incidence rate, current management options, and how your research fits into this. This approach would be perfectly acceptable, if a bit dry.

So, how might you liven it up? The subject doesn't exactly lend itself to a stand-up comedy routine. Would storytelling work though? What if you bring in the story of a family whose lives have been affected by a screening program that allowed early diagnosis and treatment ("Overcoming the Monster"), or take the viewpoint of a parent whose life has changed because of their child's illness ("Journey and Return"), or talk about the possibilities that a universal screening program might bring (Steve Jobs' "world before and world after" approach)? Any of these options will make your presentation much more memorable. You will still include all the necessary details of your work, but this time the facts are enveloped in a story that adds emotions, purpose, and a human dimension to the groundbreaking lab work that you have been doing. A sprinkle of *pathos* on your *logos* (mwah!—chef's kiss).

References

Booker, C. (2004). *The seven basic plots: Why we tell stories*. A&C Black.

Morgan, N. (2014). *Power cues: The subtle science of leading groups, persuading others, and maximizing your personal impact*. Harvard Business Press.

Protectstar Inc. (2013). iPhone 1—Steve Jobs MacWorld keynote in 2007—Full presentation, 80 mins [YouTube Video]. In *YouTube*. Retrieved March 13, 2020, from https://www.youtube.com/watch?v=VQKMoT-6XSg

Shakespeare, W. (2011). *Hamlet*. Oxford University Press Southern Africa.

Shakespeare, W., Mowat, B. A., & Werstine, P. (2020). *Henry V*. Simon & Schuster Paperbacks.

The Conversation Media Group Ltd. (2024). *The conversation: In-depth analysis, research, news and ideas from leading academics and researchers*. The Conversation. Retrieved January 14, 2020, from https://theconversation.com/europe

Planning, Preparing, and Performing Persuasive Presentations

What You Will Learn From This Chapter

Ever since antiquity, intellectuals, scholars, and scientists have held the art of giving speeches, lectures, and presentations in high esteem. Striving to be a part of this tradition will undoubtedly benefit your career. In this chapter, you will learn how to:

- develop excellent presentations while honing the skills of a great speaker and listener
- manage an audience during talks; and
- reduce stress and performance anxiety.

7.1 A Recipe for a Successful Speech

[1]Giving presentations and lectures that are worth listening to is a core skill for any researcher. You must be able to deliver presentations that carry a clear message, are easy to follow, and contain just the right amount of information. Most importantly, you need to learn how to be convincing by winning the hearts and minds of your listeners.

To ensure success, we suggest serving an Aristotelian dish (see Sect. 4.4.1): put some nutritious *logos* (food for thought) on the plate, soak it in *ethos* (believability), and add a dash of *pathos* (emotions) to finish. *Bon appétit*—now you can offer your audience something truly persuasive!

---

[1] The text of this chapter contains material which was translated from a seven-part article series in Swedish in the business magazine *Kommunicera*, published in Göteborg 2013. The translation was made by Michael Hinton, Ditchling, UK.

> **Oral Communication Tips in This Book**
> Besides this ambitious introduction here in Chap. 7, practical instructions and advice for different speaking situations can be found in both Part II and Part III of the book:
>
> - Section 6.4: Telling a Good Story
> - Section 10.2: The One-Minute Elevator Pitch
> - Section 10.3: Marketing an Idea, Project or Product
> - Section 10.4: Participating in a Panel
> - Section 10.5: Giving Longer Talks and Workshops (1–6 h)
> - Section 10.6: Planning and Executing a Teaching Session
> - Section 10.7: Turning a Live Lecture into an Online Lesson
> - Section 10.8: Giving a Media Interview
> - Section 10.9: Giving an Official Speech or Informal Toast.

7.1.1 The Power of the Spoken Word

Before discussing the practical matters of public speaking, let's reflect on the power of oral communication; it is quite extraordinary. Throughout history, humankind has invented elaborate ways to communicate on a large scale for maximum effect: aggressive propaganda and advertising campaigns, magnificent parades, bombastic rock tours, gigantic movie productions, and more. These ambitious communication activities involve huge amounts of resources—man-hours, time, money—to promote awe, impress audiences, hammer in messages, or manipulate people's behavior.

Yet, they can all be outdone by a lone but skilled speaker. Think about it: a few well-chosen words from a trustworthy person, conveyed with passion, style and sensitivity at the right time and in the right context, can affect and persuade people more effectively than any gargantuan communication activity. Speeches like the ones below all have the potential to stick with people, change the way they think, make them reconsider ideas and values, and—perhaps most importantly—promote action:

- A transformative talk from a brilliant role model in science.
- Paradigm-challenging news from an upcoming talent during a conference.
- An inspired TED presentation from an influencer.
- A farewell talk by a popular and wise professor and mentor.
- A comforting speech at a funeral.

As a matter of fact, the authors of this book dare to state that the spoken word is—in the long run—probably the most effective, convincing, and awe-inspiring communication channel of them all. Skilled speakers have the power to reach both the intellect and emotions of people in order to influence, perhaps even control, the discourse. It doesn't matter if their voice comes from a stage, from the floor in front of a whiteboard, or from the other side of a coffee table.

So never say, "It's just a presentation; who cares?" Instead, see the spoken word as something that has the potential to change the state of things, and even cause disruption.

> **Reflection VII.I**
> Some speakers radiate a special kind of energy and "follow me!" vibe from the stage. Typically, they appear as outgoing, expressive, enthusiastic, and naturally persuasive.
>
> Now, say that you happen to be a person of his kind. Do you think the Spider-Man principle ("With great power comes great responsibility") applies here? That is, should you moderate your enthusiasm in order to give people a chance to look at your messages with somewhat skeptical or critical eyes? Or is it up to them to not get swept away by your charisma?

7.1.2 Becoming a Great Speaker

Let's look closely at what characterizes a great speaker, no matter the setting. This is what you should strive for.

The Basics

Passion for Your Subject
You love your subject, and you have the ability to display this fact to the world. You know that your enthusiasm is contagious, and you are not afraid to transfer this contagion to people who cross your path.

Clarity
You have the ability to convey the essence of your topic and present it in a structured way that is easy to follow, understand, and remember.

Authenticity
The messages and values you convey correspond to what you represent as a person. You are not interested in hiding away your true self or what you believe in.

Presence and Responsiveness
You are present in the moment, fully focused on your audience, their questions, and their needs. You respond aptly to what happens in front of you (the audience) and around you (the context).

Professional Attitude

You come to the event well-prepared, ensure people can hear and understand you, and treat everyone with politeness and respect.

Helpful Talents and Skills

Empathy

If you want to develop an interactive style as a speaker, going beyond professionalism and politeness, you need to understand the emotions, motivations, and views of other people—you have to be able "to walk a mile in their shoes."[2]

Clear, Strong Voice

Some are born with voices that are easier to listen to than others. However, everyone can improve their voice skills by learning the right techniques and practicing. Skilled speakers know how to keep the muscles affecting the vocal organs relaxed so that their voice doesn't get strained and tired after a few hours (see the box in Sect. 7.3.1 for guidelines on how to look after your voice).

Solid English

A "foreign" accent is not a problem, as long as it doesn't result in intonation or language sounds that deviate considerably from "standard English" (whatever that is!). Never apologize for your accent—it's part of who you are. If you worry that the audience might spend time wondering where you are from instead of focusing on your talk, you could drop a small remark along the lines of "and this accent comes to you all the way from Chile."

However, if your friends note that your English is hard to understand, it is probably worthwhile to find a language coach; it doesn't have to take long to identify where your pronunciation falters and to help you fix it.

Here's a special note to native speakers who are speaking English in a local dialect or accent. This may come as a surprise, but your way of talking may actually be very hard to follow for people who use English as a second language—even if they speak it fluently themselves. This is especially apparent if you use colloquial language mixed with slang.

Humor

As a speaker, you find yourself in a particularly exposed position; you're on your own on the podium, whereas the individuals in the audience have the benefit of safety in numbers. In this asymmetrical power situation, humor can be your best and truest friend. In the hands of a person with social sensitivity, humor is a great way to create connections, make people relax, and smooth out social friction. However, if you feel

---

[2] Originally from the poem *Judge Softly* (1895) by the American writer Mary T. Lathrap: "Just walk a mile in his moccasins/Before you abuse, criticize and accuse."

that you don't have comedic talents, or if you are not really that good with people, it is ok just to be serious, clear, and well-structured.

Charisma
This trait is difficult to define and describe. Some people—and it is often hard to spot who they are beforehand—just have a special presence on stage that pulls in the audience. On the one hand, they may be popular individuals with a leading social position, wit, and very apparent humor. On the other hand, they may also be people who don't make much fuss about themselves; they just go through a wondrous personality transformation when they enter the limelight.

All this said, you should never feel forced into a speaking style which isn't "you." The authors of this book strongly believe in diversity when it comes to the art of speech. If speakers were musicians, we'd appreciate the disciplined but subtle pianist playing Bach preludes just as much as the flamboyant soul diva singing "Halo" or the tongue-in-cheek singer-songwriter performing rhymed stories.

Reflection VII.I
What is your position in the graph below? What kind of speakers do you like to listen to? (Fig. 7.1).

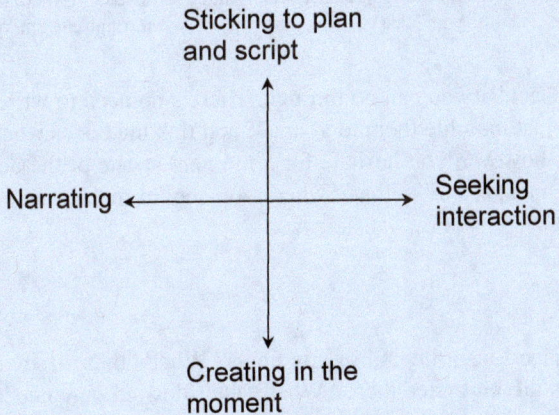

Fig. 7.1 The speaker spectrum—where do you fit?

7.2 General Preparations for Any Talk

Step 1: Analyze the task

As with all communication activities, the first thing to do is to ensure you know what you signed up for, and what is expected of you. Remember, it is not always you who sets the goals. In many situations, it is the people who invited you, based on the needs of the audience you serve—see Table 7.1 for examples.

Table 7.1 Examples of a range of communication roles and aims

Your role	Who sets the goals?	How to proceed
Conference presenter	**You and the organizer**. It is expected that you contribute to the intellectual spirit of the conference, strike a positive note, and share interesting information, interpretations and opinions	Let your expertise, experience, and passion for the subject serve the audience and those who invited you
Invited tutor	**The organizer**. The faculty's teachers, students and administrators expect you to follow—or at least complement—the curriculum and the brief	Decide on the content and the schedule in detail with the organizer during the preparations. Stick to the plan
Influencer, guru or an inspirational speaker	**You**. It is expected that you should represent yourself and your ideas	Say what you want and send the invoice! (Alternatively, escape through the emergency exit)

Below is a checklist you can go through. There's no need to write long answers to each point—just mumble them to yourself and tick the boxes where the answers are obvious. If, however, you hesitate for a moment at one of the points and think "That's a little unexpected!"—that's where you should make a note.

Checklist

Whys

☐ Why have I been invited by **my hosts**? What's their desired outcome?
☐ Why is the **audience** there? What's their desired outcome?
☐ Why am **I** here myself? What's my desired outcome?

Examples of desired outcomes: receive/convey/exchange information or skills, be inspired/entertained, learn/teach, sell/market, manage/lead, give and take feedback/critique, discuss/debate, find common ground/get someone to change their opinion.

7.2 General Preparations for Any Talk

The audience
- ☐ What is the social and professional profile of my audience?
- ☐ How familiar are they with my subject? What is their (current) attitude towards it?
- ☐ What signals will help to build their trust?
- ☐ What will engage them?
- ☐ Which factual arguments will persuade them?

Presentation
- ☐ What presentation methods can I use?
- ☐ How shall I manage my content?
- ☐ What can and should I bring with me?
- ☐ How shall I structure my talk?
- ☐ How shall I present it?
- ☐ Is it appropriate to have a gimmick of some sort?

Interaction
- ☐ Can I and should I interact with the audience?
- ☐ How do I finish? On what note?
- ☐ How do I get feedback?

Setting
- ☐ Where will I be presenting? What is the set up of the room?
- ☐ What are the limitations (acoustics, size, unmovable furniture)?
- ☐ What are the resources available (portable microphone, laptop, whiteboard…)?

Step 2: Clarify Your Objectives

Now that you know what the task involves, it's time to revisit your objectives. You probably did this fairly quickly at first, but now ease the tempo and really reflect.

The American public speaking coach Nick Morgan once said, "The only reason to give a speech is to change the world."[3] So—how do you intend to change the world with your presentation? (See also Sect. 6.2 for more reflections on setting goals.)

- **Do you want to communicate information and knowledge?** If so, what do you want your audience to do with this information? Should they simply acknowledge it, or are you seeking some kind of critique or opposition? Do you want them to change their behavior, or make specific judgements or decisions based on their newly acquired knowledge?
- **Do you want to sell something?** If so, are you aiming for short-term sales (where buyers decide on the spot, here and now) or long-term sales (where buyers go home, think about it, and maybe decide to buy in a few months' time)?

---

[3] Morgan (2014).

- **Do you want to lead and motivate a team?** If so, what steps should you take to gain and/or maintain their trust?
- **Do you want to change their minds?** If so, what's the starting point? Are your listeners easy to reach, or have they entrenched themselves behind fixed opinions? And how much prestige is at stake—can they change their minds without damaging their self-image or altering their social status?

Step 3: Compile Your Content

Now it's time to gather the main messages, creative ideas, and supporting information you want to add to your slides.

1. **Write a brief summary of your talk**.
   You might have already done this if you sent in a conference abstract or a pitch to the event organizers.
2. **Define your three main messages**
   If your audience were to forget everything from your presentation except for three statements, what should these three statements be?
3. **Compile your proof**
   This includes graphs, images, explanatory graphics, examples, arguments, and quotes that support your main messages.
4. **Decide on a persona**
   How should you present yourself and your activities to the audience to appear trustworthy and interesting?
5. **Invent a hook**
   Is there something that you can use to create a strong start that gets listeners to prick up their ears? Photos, scanned news articles, social media screenshots, short video clips or quotes are especially useful for this.
6. **Look up a fitting story**
   Consider whether you have a good story up your sleeve; something that can elevate your presentation, like the lightbulb moment of the research idea, a stroke of luck during tedious lab work, a funny field study incident, an unexpected cross-faculty partnership, et cetera.
7. **Decide if you want to finish with a "call to action"**
   If so, make the instructions clear and ensure the audience knows what the next steps are.
8. **Reflect on ethos, pathos, logos**
   In addition to the list above, consider how you can use the three means of persuasion for the best possible outcome (see Sect. 4.4.1).

Step 4: Plan and Structure Your Presentation

Now that you have identified all the material that could be interesting for your presentation, the next step is to create a structure. The best way to do this is to use sticky notes (or their digital equivalent, if you prefer).

7.2 General Preparations for Any Talk

1. **Write everything down on sticky notes**.
   Put one thing you could consider including in your presentation on each sticky note. Place them on an empty table or a whiteboard.
2. **Organize the sticky notes in clusters**.
   You will note that a "sticky note mind map" starts to take shape before you. Normally, things fall into place quite quickly. Don't dwell on the details—there's time for that in step 4.
3. **Make a "storyboard" of your slide deck**.
   Now, start arranging the sticky notes in well-ordered columns. Let each note correspond to either a PowerPoint slide, or something you should demonstrate or perform outside the slide show, such as showing a sample or telling a story.
4. **Incubate your structure**.
   Let your sticky note structure stay in place on its table or whiteboard for a couple of days. Look at it every now and then to add, remove, and modify items.

At this point, pick one of two strategies. If you are the type of speaker who likes to add a (relevant!) anecdote on a whim or does not want to be rushed by any technical problems that might arise, apply the **two-thirds rule:** when you are finished, you should have material that will last about two-thirds of the time you have available (so, if you have been given a ten minute slot, prepare material for seven minutes). If, however, you know that you are likely to speak very quickly and don't want to risk finishing in half the allocated time, apply the **120% rule**: have an extra 20% of "bonus material" tucked away after the last slide, which you can then bring out during the Q&A session. In all likelihood, you will not need to use it, but it will give you some peace of mind.

Step 5: Produce the Slides

At last, it's time to switch on the computer and create visuals for your presentation. This will take less time because you've already thought through everything in detail, collected the material, and drafted a clear structure. For more tips on making slide decks, see Sect. 12.2.

During this stage, consider the following:

- Strive to make the slides as simple as possible—each one should contain only one message or one strand of thought.
- Ensure the background of the visuals is a single color. Remove dates, multiple logos, and other rubbish.
- Make all illustrations BIG. Stretch photos from edge to edge. Make sure the captions on figure axes are legible.
- Use a font size of 36 points or larger. This not only makes the text easier to read, but also prevents you from squeezing too much of it onto each slide.

- Ration the number of bulleted lists. They are perfectly ok for overviews and summaries, but their exaggerated use is also largely responsible for the phenomenon of "Death by PowerPoint." To avoid boring your audience to their demise, our rule of thumb is that no more than 25% of the slides should contain bullet lists.

Step 6: Add Finishing Touches and Simplify

Once your slide deck is ready, leave it for a few days if possible. When you return, put on your critic's spectacles and your editor's hat and review the material. Be strict with yourself and edit according to these guidelines:

- Focus on the main messages—remove content that belongs to another story ("Kill your darlings![4]").
- Use fewer words.
- Use fewer bullet point lists.
- Use fewer numbers.
- Include more photos and clear visuals.
- Simplify charts.

Step 7: Test, Rework, and Practice

Now print your slides as thumbnails (we suggest nine per page)—this will be your script. Then go to a conference room, start a timer and go through your presentation. If you don't manage to keep within your time limit, remove some content. You will probably also notice that some parts of your presentation do not work in practice in the way you intended.

If you have problems memorizing all the details of your material, either use the presenter mode on your computer (check that it works, and how to troubleshoot it if it doesn't), or prepare some cue cards in A6 format (105 × 148 mm), like the ones television hosts use. Speaker's notes definitely make you feel more secure if you anticipate being nervous before your talk, and preparing them will help you memorize your content.

When you're beginning to know your material well, practice with a trusted friend who gives honest but kind feedback[5] (some guidelines on giving feedback are provided in Sect. 8.3.2). It is particularly useful to pinpoint which parts are

---

[4] Originally from a Cambridge lecture series, *On the Art of Writing*, by the British author Sir Arthur Thomas Quiller-Couch: "If you here require a practical rule of me, I will present you with this: 'Whenever you feel an impulse to perpetrate a piece of exceptionally fine writing, obey it—wholeheartedly—and delete it before sending your manuscript to press. Murder your darlings."

[5] Joanna wants to make a note here that British people are, unfortunately, useless at this. Because they are very polite and hate confrontation, they will usually say something along the lines of: "It was lovely" (which isn't great for your growth), or "That was interesting" (which is code for "terrible," but you might miss it if you're from a different culture). For direct and to-the-point feedback get a Dutch friend instead; the Dutch are famously blunt. You might leave scarred, but you'll definitely have loads to work with.

not clear or engaging. If possible, practice in an environment similar to the place where you will give your presentation. A video recording may be helpful, but only for those who are comfortable seeing themselves on screen. (If you're not, leave it be! Our experience is that video recording does more harm than good in these cases.)

At this late stage, it is simplest to follow the principle "when in doubt, leave it out" and delete anything that is questionable. Revisit the goals and main messages you defined to see what absolutely must be retained and what can be omitted. When you feel fluent enough, find a friendly colleague who can offer encouragement, a few wise comments, and some good advice along the way—and try your talk out on them.

Step 8: Testing and Preparing the Setup
Connect all equipment and ensure you have signal. Turn on the remote, open the slide deck, and view a few of the slides. Walk around the room to see how the screen looks from different spots in the audience. Make sure there is water available. Identify any tripping hazards.

Step 9: Change the World!
The stage is yours—take a deep breath, relax and have fun! It's ok to bring your cue cards on stage if you need them; enjoy the feeling of acting like a television host.

7.3 Presentation Delivery

Providing written instructions on how to perform during a presentation is somewhat futile—it's like learning to paint watercolors or play the harmonica from a book.[6] However, we'll do our best to convey some useful advice.

A great starting point is the TED Talk presentation "The Magic Washing Machine" by Professor Hans Rosling from 2010[7]—in our eyes, one of the most complete examples of the tools and techniques a modern speaker can use to get a message across. The video is available on ted.com. While you watch it, observe the following:

- Professor Rosling has a modest personal image, radiating wisdom and long experience, rather than ambition and power.
- He moves slowly across the stage between different stations where different parts of the talk take place.
- He uses gestures, but does not overdo it.

---

[6] Believe it or not, Olle actually learned to play the blues harmonica from a book he bought in a music store. This was twenty years before YouTube.
[7] Rosling (2010, December).

- He speaks English with a strong Swedish accent, but he doesn't try to moderate it; he feels secure in his identity as an enthusiastic professor from Scandinavia with something important to tell the world.

Please note that the TED Talk format—giving a prepared speech from a stage—makes it hard to have close interaction with the audience. This comes much more naturally when standing on the same floor level in a normal conference room.

> **Reflection VII.II**
> Have a think about Hans Rosling's TED talk mentioned above.
>
> - What would you say his main messages are?
> - How did he use the rhetoric triad of ethos, pathos and logos?
> - How did he use storytelling, and did it work with you?

7.3.1 Using Your Body to Communicate

During a talk, you can use your body in a variety of ways. Here are some suggestions—with the disclaimer that they are, of course, suggestions; tailor your performance to your personal style and what you feel comfortable with.

Voice
As a speaker, you can do many interesting things with your voice. Besides pausing, you can change the pace and pitch, and modulate how carefully you articulate words. Some speakers repeat important phrases to help the audience remember them. One of the most important details is to learn to emphasize the right places and vary your tone. A uniformly quick presentation is just as boring to listen to as a uniformly slow one.

Always adjust your volume to the size of the room—don't deafen the audience with a booming voice in a small seminar room, and if you are in a large auditorium, use a microphone. Additionally, one of the most effective ways to capture your audience's attention is to lower your voice to a whisper (if the setup permits). This is often far more effective than raising your voice.

Use Silence
Many presenters are terrified of silence and fill gaps with "err," "like," "umm" and other non-words. Yet silence can work to your advantage. A brief pause to gather your thoughts helps the audience digest what has just been said. You can also use a pause for emphasis—an important point followed by silence gives it more gravitas and makes the audience pay attention.

Look After Your Voice
Avoid shouting in noisy rooms, and take it easy during that football game or rock concert. Every time you become hoarse, you strain your vocal cords; the more often it happens, the higher the risk of permanent damage.

The importance of your voice becomes evident the day you cannot use it. Experienced speakers can manage fever, headache, back pain—some painkillers and strong coffee, and they're ready to serve their audience (and collapse afterwards). But if they lose their voice, they're helpless—no matter how important the occasion.

> **Seven Ways to Take Care of Your Voice**
> *By Lydia Flock, a voice coach, voice researcher, and vocal massage therapist; https://www.oxfordvocalmassage.co.uk/.*
>
> **Tip 1: Warm up your voice**
> Just like it is important to warm up before you go for a run or lift weights, it is important to warm up your voice—ideally daily, but especially before singing or speaking for an extended period. A strategic warm up can make a huge difference in not only how you sound (e.g., resonance and tone), but also in taking care of your vocal health. A 5–15 min warm-up can make a big difference. If you're unsure where to start, see the exercise box below.
>
> **Tip 2: Cool down your voice**
> I like to think of a vocal cool down as 'closing up shop' at the end of the day. Often forgotten and neglected, taking some time after an extended period of vocalizing to do a few descending lip trills, hums and stretches can really help manage a busy vocal lifestyle and set your voice up for success the next day.
>
> **Tip 3: Hydrate**
> Our vocal folds (sometimes called 'vocal cords') are a mucous membrane. They need lubrication to come together to make sound with ease—super important when considering our vocal stamina! Aim to drink approximately 2000–2500 mL of water a day, based on drinking 1 mL for every calorie burned. Drinking little and often is more advisable than drinking a lot infrequently. If you are rehearsing, performing, or exercising, add an extra 100–150 mL per hour worked to this number.
>
> **Tip 4: De-stress**
> I often say, 'Self-care is voice-care'! That's because stress can affect the voice in millions of ways. Not only can stress make it physically harder to use our voices (when we get stressed, we often hold more tension in our bodies), but it can also make the emotional experience of using our voices

less fun. So, take time for some self-care each week, whatever that may mean to you.

Tip 5: Take voice naps

Short periods of voice rest—or 'voice naps'—can be an effective strategy to combat vocal fatigue, particularly if you have a heavy day of using your voice (for example, back-to-back lectures one afternoon). How often and how long should your voice naps be? It really depends on your schedule, but one good rule of thumb is that for every 40 min of speaking, aim for 20 min of voice rest.

Tip 6: Work with a coach

Whether you are a certified vocal health nerd or you have no idea how to begin looking after your voice, working with a coach can help you achieve your vocal goals. You may want to increase your stamina after speaking for a long period, or improve your vocal range and resonance—whatever your objectives, seeking feedback is a crucial part of a good voice practice. It can be tempting to want to figure out everything you need for your voice on your own—and yes, you can learn a lot from reading online—but sharing your goals and challenges with a vocal coach can help bring you closer to these goals…and get you there faster!

Tip 7: Get a vocal massage

Research on vocal massage is ever-growing, and suggests it is a potentially vital tool for elite singers and professional voice users. Whether you are aware of voice-related tensions or not, going for a monthly maintenance treatment can help optimize vocal health, flexibility and ease. If you can't attend a treatment in-person, you may be able to learn self-guided techniques online to care for your voice anywhere in the world.

There are many tips I left out (like using a nebuliser and straw phonation) so please feel free to do more reading on the topic if it piques your interest.

Exercise VII.I The 5 x 5 x 5 Voice Warm Up

This incredibly short warm up can be repeated throughout the day for a voice reset. It includes three exercises, repeated five times each (hence 5 × 5 × 5). All three exercises can be done in a variety of ways—on a hum, lip trill, tongue trill (sometimes called 'raspberries'), through a straw—the possibilities are endless. For beginners, start with a hum. As you voice, try to maintain a relaxed posture and a slight 'yawn' posture (this drops your larynx—colloquially referred to as your 'voice box'—slightly).

Exercise 1: Voice [on a hum/lip trill/tongue trill/through a straw] a low-ish note for 3–5 s (a note you might typically speak at). Repeat five times.

> **Exercise 2**: Voice [on a hum/lip trill/tongue trill/through a straw] a high to a low note. This sounds like an exaggerated sigh of relief and should be approximately 3–5 s, too. Repeat five times.
>
> **Exercise 3**: Voice [on a hum/lip trill/tongue trill/through a straw] a low to high to low note. This can be a relatively small range of notes—it does not need to be too high or too low and should take about 3–5 s total. Repeat five times.
>
> By Lydia Flock, a voice coach, voice researcher, and vocal massage therapist[8] https://www.oxfordvocalmassage.co.uk/.

Eyes

They say that eyes are the window to our soul. It is amazing what the eyes and the surrounding facial expressions can convey—so subtle, so fast.

Keep Eye Contact

Scan the eyes of the participants and continuously read their reactions; this gives you important feedback. Whether they approve or disapprove, whether they follow you or feel estranged—everything can be read in their eyes. Try to maintain eye contact with your entire audience (not necessarily at the same time—quite impossible in a large lecture theater); spend a few seconds looking at different areas of the room. When speaking in a bigger room, look at the people at the back of the audience—this will make the front rows feel included; on the other hand, focusing on the front row will exclude the people further back.

However, avoid staring at anyone for too long, as this will probably make them uncomfortable. Equally, if keeping eye contact is uncomfortable for you, try focusing not directly on the eyes, but something nearby—the mouth, or a point near someone's head. Or pan the room without focusing on individual faces.

Face and Head

With your face, you can either show your mood and reactions in an authentic way or in a humorous, exaggerated way. Avoid trying to do something in between, as you're at risk of either looking manipulative or corny. Nodding when you agree or to confirm that you listen helps connect with the audience.

Posture

Your posture can be very revealing. When we're feeling insecure and out of place, it is easy to literally shrink. Conversely, the best way to look confident is to stand tall, stay grounded and own your space.

---

[8] Flock (2021).

> **Exercise VII.II Practice Your Stance**
> If you want to practice improving your speaking stance, here are some pointers—do as many as you are comfortable with and capable of.
>
> - **Stand tall**: keep your head high, shoulders down, and spine long (imagine a string attached to your skull, pulling you up until your backbone is nice and elongated).
> - **Stay grounded**: stand with your feet about hip-width apart and knees relaxed for better balance. Wear comfortable shoes: sky-high heels might make you look fabulous, but they will also push your center of mass forward, affecting how comfortably you breathe; they're also trickier to balance in.
> - **Breathe**: let your belly flop out a bit to allow your diaphragm to work when projecting your voice. Relax your shoulders—it is worth rolling them a few times before the presentation to expand the chest and let more air into your lungs.

Hand and Arm Gestures

Your hands and arms are the perfect props—they can underline key points, accentuate different sections of your presentation, and help connect you to your audience.

Some speakers feel very conscious about their hands and comfort themselves with fidgety actions, such as playing with their rings, clicking a pen they are holding, or "petting the hamster"—stroking one hand with another as if holding an invisible pet. If you feel the need to hold something (akin to a security blanket), use the remote—it fits in your hand, is useful, doesn't make annoying sounds, and the audience expects you to hold it anyway. It's the perfect alternative to that poor, invisible hamster.

If your hands tend to get a bit shaky when you're stressed, avoid notes written on A4 paper—these large sheets will amplify the shaking, and distract both you and the audience. Instead, make notes on small sheets of thicker card that fit easily into your hand and don't flap around.

> **The Box Box**
> Think about adjusting your gestures to the size of the room. What does that mean? Imagine that you move your hands within an area delimited by a box (Fig. 7.2). Try to keep this box at chest height—somewhere between your neck and your waist.

7.3 Presentation Delivery

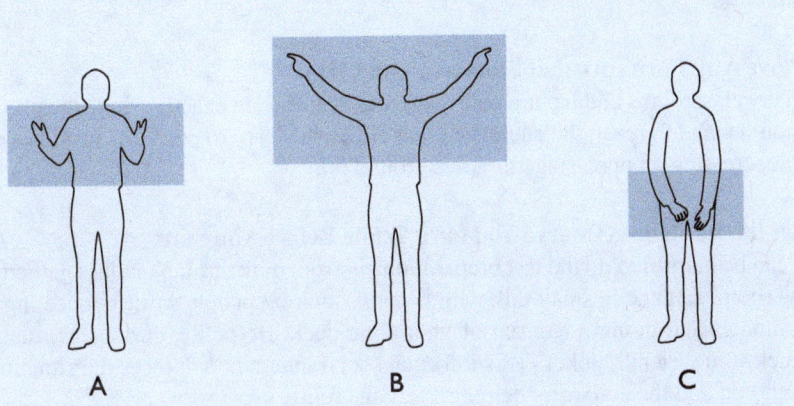

Fig. 7.2 Think about the placement of your gestures. When speaking in a regular-sized lecture theater, aim for chest height (**a**), to avoid seeming over-the-top (**b**) or overly timid (**c**)

If you lift the gesture box above your head, your body language will look too big for a regular auditorium, though would not be out of place at a concert or an Olympic opening ceremony ("Welcome, world!"). If your gesture box drops to around your hip height, it will make you look insecure and shy.

Project Your Enthusiasm Towards Your Audience

From the moment you begin, show that you feel comfortable and love your subject. The best talks are about having a good time together with the audience.

Still, there will be times when, through no fault of your own, you find yourself speaking to an audience with low energy—perhaps your presentation is right after lunch, first thing on a Monday morning, or at the end of a long day. Talking to people who are visibly tired or unfocused can be draining, because it requires you to increase your energy as a speaker. If the presentation is particularly important, you can recruit a friend to assist you: plant them in the middle of the auditorium and ask them to smile widely and nod vigorously at whatever you are saying. Such a positive response, even from a single person, will energize you—and your energy will, in turn, lift the rest of your audience.

7.3.2 Using the Space Around You to Communicate

Getting to know your presenting space will let you feel more at ease on the day, and troubleshoot any problems ahead of time. Try to familiarize yourself not just with the presenter's view, but also walk through the audience seats to learn about their perspective.

The Stage

Move Any Furniture that Stands in Your Way
Many classrooms and lecture halls have sturdy furniture exactly where the presenter should stand to reach the audience most effectively. Try to push this furniture to the side, creating an open stage to move around on.

Let the Audience Observe You for a While Before You Start
It can be helpful to do the last preparations in front of the group or class gathering in the room. Engage in small talk with friendly-looking people while connecting your computer and doing a test run of your slide deck. Be polite, curious, cordial, and crack some friendly jokes. This informal social interaction defuses the situation for both you and the audience, helping everyone relax.

Use the Space on the Floor Purposefully

- Move slowly from left to right and back again, then stand still for a while. Repeat. This helps keep the audience alert and gives you a better overview of all the faces looking at you. Remember: no pacing like a stressed tiger in a cage—keep it nice and slow!

  Note: if you know there are hearing-impaired audience members, try not to turn away from them; make sure your face is visible to help them follow your talk.

- Step forward to emphasize points. Close in on the audience when you want to stress something, ask a rhetorical question, or create tension in the room. Move back when you start lecturing again, but do so after a brief pause to signpost to the audience that the lecture is reverting to its previous pace (Fig. 7.3).

Fig. 7.3 Consider using the left-to-right movement on the stage for a bit of visual distraction; the forward-backward movement, however, is a powerful tool for emphasizing points or building a stronger relationship with the audience

- If you want to have a discussion, you can pull out a chair and sit down so that you become one of the group.

The Technology

Use a Remote
While some venues may provide a remote, it's best to bring your own. The usefulness of a remote cannot be overstated: almost everything on the list below becomes easier when you're not tethered to your computer. Plus, as mentioned above, holding something in your hand soothes the nerves a bit.

Use a Microphone
If your voice is naturally soft and you're speaking in a large auditorium, don't try to scream—it's exhausting and unhealthy; ask for a microphone instead. You might be given a lapel microphone, which clips into your top, and a body-pack containing the transmitter and batteries (a small box that clips to your belt or fits in your pocket). Joanna feels obliged to highlight that these microphones are not designed with women's fashion in mind—if you are wearing a dress with no belt or pockets, you will need to hold the body-pack in your hand. Equally, if the microphone is attached to an ear-piece, be cautious of dangly earrings that might knock on it and create noise.

If you are given a hand-held microphone, do not gesture with the arm that holds it—the constant volume change will distract both you and your audience. The most secure way of managing a hand-held mic is to rest it gently on your chin and keep it there until the end of the talk. When it's "glued" to you in this way, you're less likely to wave the microphone-bearing arm around (though by all means gesture with the free one).

If the only microphone available in the room is permanently attached to the lectern—run[9]!

Use the "B" Key
Microsoft PowerPoint, Apple Keynote, and Google Slides all share a great shortcut: pressing the "B" key on the keyboard replaces the current slide with a black screen. This action shifts the audience's attention from the slide deck, and directs it towards you. Now they are ready for the next part—you have their attention, so offer them something interesting. (From the speaker's point of view the effect of using this command is almost entertaining: show the slide, and they look at the slide; show the black screen, and they look at you.)

Pressing the "B" key is also a polite and humble way to respond to audience questions. By removing your current slide from sight, you invite interaction and show that you are fully focused on their question. However, be aware that pressing

---

[9] ...towards the feedback form, where you politely suggest some improvements to the facility manager.

the "B" key on French keyboards will show a white screen, as it stands for "blanc," not "black."

Creating Dynamics
Never let yourself fall into a lull. Instead, keep working hard to make your presentation interesting to listen to. Change the pace, ask questions, engage in discussions, allow for moments of silence—stay dynamic! When you are about to discuss a particularly interesting slide, don't show it immediately; instead, create anticipation by holding it back for a few extra seconds.

Switch Between Modes
Do your best to create variety. Show some slides, run a Mentimeter survey, switch to a whiteboard, and then distribute an exercise on paper. The audience's attention span is short—make it easy for everyone to stay engaged.

Do Something Unexpected
If your self-esteem is up for it, try following an impulse that pops into your head during a talk. To allow a crazy idea past your self-censorship, it must meet the following criteria:

- Its purpose should be to give the audience a more memorable experience—not to boost your ego.
- It should be clear *why* you're doing it. Avoid surrealistic stuff that brings confusion rather than clarity.
- It should only involve people who feel safe within themselves; while it may be ok to embarrass an experienced colleague, it is not acceptable to involve a student in something they have not agreed to.

Improvise
The best thing about being well-prepared is that it sets the scene for successful improvisation. If you know your material, yourself, the audience, and are in control of the time schedule (by having practiced and by applying the two-thirds rule of Sect. 7.2) you can allow yourself to act on ideas that come up in the heat of the moment. Most professional speakers, actors, musicians and comedians have anecdotes about how magic was created between them and the audience, totally unplanned.

7.4 Making Everyone Feel Safe

The world changes, and what's considered inappropriate, offensive, or an intrusion of personal space today is not the same as say, thirty years ago. Still, many speakers—not least senior ones—often forget to address this.

If you feel that you are navigating unknown waters when it comes to themes like gender, race, historical privilege, etc., it may be advisable to refrain from discussing them if they are not included in the lecture or seminar agenda. Some general advice: don't make jokes about gender roles, don't refer to cultural stereotypes, and proceed with caution whenever you approach the personal sphere and background of individuals in the audience.

Be mindful of your audience's needs—try to find out in advance if anyone has particular requirements, such as using a special microphone for individuals with hearing difficulties, or reducing color contrast on your slides for a neurodiverse public.

7.5 Preparing for a Q&A

"It ain't over till it's over"—especially for science presentations where the rehearsed, lecturing part is in most cases followed by questions. In fact, the first part may just be a warm-up, for example during a prestigious conference or thesis defense.

You will never be able to foresee exactly what questions you will encounter during a Q&A. You can, however, prepare yourself in a way that makes the process much smoother, at least compared to relying solely on thinking on your feet.

1. Engage some colleagues and friends in a brainstorm

Gather one to three friends, sit down together with a stack of Post-its, and decide who is going to take notes—one question per Post-it. Then, try to penetrate the subject from all possible angles. If you are preparing a talk for a non-expert audience, include people who don't share your expertise and views.

When you're done, organize the sticky notes on a table or a whiteboard (just like you did earlier here in Sect. 7.2) and take a photograph.

2. Write replies using the ethos, pathos, logos model

During a scientific presentation, it is generally expected that you present the facts. However, after studying the section about rhetoric (see Sect. 4.3), you will realize that many questions actually need an approach that focuses on credibility, shared opinion and values, and sometimes even feelings and emotions. Keep this in mind when preparing responses to the themes on your Post-it notes.

3. Engage a *Advocatus Diaboli*—a Devil's advocate—during rehearsal

The *Advocatus Diaboli* was an official position in the Catholic Church during medieval times. In the process of canonization, where the church discussed whether a pious person should become a saint, one official had to represent the view of Satan himself. Their task was to see everything from the most cynical, pessimistic, and conspiratorial perspective.

Today, a devil's advocate is someone who engages in a debate, "arguing against something without actually being committed to the contrary view."[10] In this case, let one of your friends play the role of a nightmare opponent, giving you the strongest resistance possible during your rehearsal.

7.6 Managing Your Nerves

As a communications trainer, one of us (Olle) has coached hundreds of STEM students at Bachelor's, Master, and Ph.D. levels, as well as young postdocs. According to (anecdotal) observations made during these events, stage fright or performance anxiety seems to be a very common phenomenon in this group—often associated with a personality type characterized by perfectionism and self-consciousness. Or, to put it more bluntly: high achievers often go hard on themselves.[11]

Therefore, let's explore the concept of performance anxiety. It refers to the stressful, emotional experience of doubting one's ability to perform a task or dreading the discomfort, suffering, or pain associated with it.

As our readers probably know, fear, anxiety, and high stress levels tend to trigger a *Fight*, *Flight*, or *Freeze* (FFF) reaction. These three Fs represent evolutionary survival behaviors that, while not particularly sophisticated, can be very effective in the right circumstances. When ancestral humans faced danger—such as encountering a threatening animal, a strong tide, or a bushfire—these reactions made sense. Increased heart rate and skeletal muscle trembling created focus and readiness to act. A passive and fatalistic reaction saved energy and avoided triggering a hunting instinct. Sometimes two additional Fs are added to the mix: *Faint*, where stress causes you to pass out, and *Fawn*, where you become subservient and servile in front of an intimidating person. These were useful, automated strategies for our ancestors, and we are living proof of their effectiveness. But in our modern world and professional lives, many aspects of our stress reactions—be they physiological, cognitive, or emotional—have become counterproductive.

So, how should we deal with all this when it hits us as performance anxiety—in an era where the predator is replaced by an important presentation?

On one hand, these reactions create chaos. When tricky situations appear in our professional lives, we typically need:

- MORE cognitive capacity for memory, analysis, and planning,
- CLEARER notion of the complexity of the world around us,
- EXTENDED sensitivity, empathy, and ability to cooperate,
- LESS activity in our stomach, bladder, and sweat glands.

---

[10] Wikipedia Contributors. (2019, April 22). *Devil's advocate*. Wikipedia; Wikimedia Foundation. Retrieved April 20, 2020, from https://en.wikipedia.org/wiki/Devil%27s_advocate

[11] This brings a Bible verse to mind: "For in much wisdom is much grief: and he that increaseth knowledge increaseth sorrow" (Ecclesiastes 1:18, King James' Version).

7.6 Managing Your Nerves

In these respects, we are apparently let down by our brains, bodies, and evolution.

On the other hand, some of the reactions—if moderate and controlled—can actually help us and boost our abilities. They provide us with:

- SHARPER focus on what is in front of us, making it easier to distinguish the essential from the trivial and to stay in the moment. If you're lucky, you might even experience the wondrous feeling of flow.
- STRONGER pizzazz, defined as an attractive combination of vitality and glamor.
- INCREASED emotional engagement; if you've ever given a talk with a voice trembling from passion or indignation, you've likely felt the communicative power of strong emotion.

In fact, it is very hard to shine on stage without the help of a controlled stress reaction.

In sports psychology, there's a classic graph, showing an "arc of tension" (see Fig. 7.4). If you don't feel any stress, you will not perform at your best. Neither will you if you panic. Instead, you should aim for a sweet spot on the curve where you optimize your abilities for the task at hand.

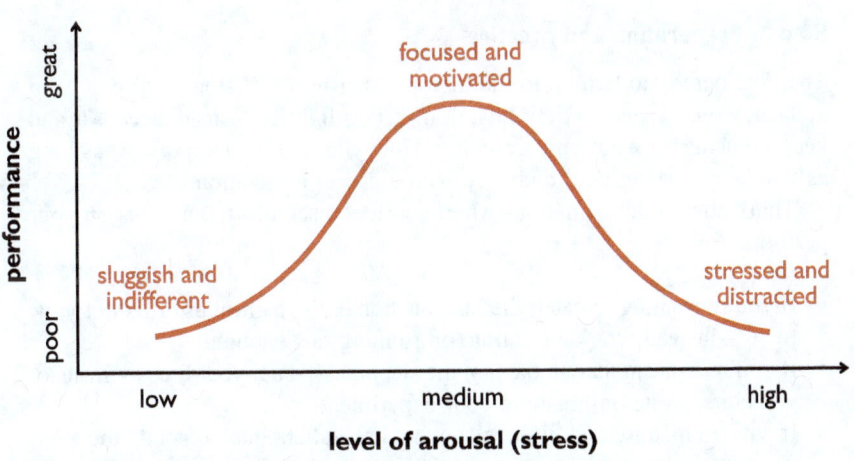

Fig. 7.4 The Yerkes-Dodson curve. The left side of the curve represents low arousal, or stress. The right side represents high arousal. And at the center is a medium level of arousal. The vertical axis goes from poor performance (at the bottom) to peak performance (at the top)

A CBT-Oriented Approach to Managing Performance Anxiety

Modern sports psychology often uses a Cognitive Behavioral Therapy (CBT)-oriented approach to help athletes perform. This includes applying the final steps of the plan they have followed for months and years, turning their preparations into controlled action. This way of thinking can also be useful for scientists giving an important presentation.

Step 1: Acceptance

The first step is to start learning how to accept reality: not shying away from discomfort, the feeling of being evaluated, and the risk of failure.

- Your presentation and its preparations will likely be associated with discomfort in the form of tension, stress, and anxiety. This is reality. Don't try to evade it; instead, learn to deal with it!
- During the talk, there may be people who evaluate you and your performance, and put you to the test by asking critical questions. This is reality. Don't try to evade it; instead, learn to deal with it!
- You may succeed (the most probable outcome, considering your careful preparations) or you may fail (a less likely option). Both outcomes are possible. This is reality. Don't try to evade it; instead, learn to deal with it!

Step 2: Preparation and practice

The goal here is to live up to the famous British WWII slogan: "Keep calm and carry on." When anxiety hits, don't try to fight it. Instead, accept it and keep focusing on what you are doing. How you feel is unimportant, as long as you keep serving your audience with a great presentation.

Think about other situations when you feel discomfort, but focus on your performance:

- If you're running to catch the last bus home, the painful effort is not your focus—instead, you concentrate on running fast enough.
- If you're working late in the lab, the sleepiness is not your focus—instead, you concentrate on finishing your experiment.
- If you help a scared little girl who has fallen into a pond, the cold water doesn't bother you—instead, you concentrate on getting the child to safety.

You can even start studying what pain and discomfort do to you. For example, observe a mosquito while it is sucking blood from your arm, or explore how it feels to become soaking wet in a rainstorm. A CBT-based exercise used in elite sports is timed bathing in cold water: participants calmly talk with the coach about the physiological reactions they experience while they patiently wait for the time to run out.

Step 3: Executing your plan

When the day comes, you will know and understand more about your stress reaction—something that helps you focus on your performance, not your feelings. An extra bonus is that the more willing you are to accept performance anxiety, the less of it you will probably feel when it is time. Acceptance is key: much of the discomfort is actually created when you are trying to fight the feeling.

This text is based on communication with Cecilia Åkesdotter, mental coach of the Swedish national team in orienteering.

7.6.1 Practical Tips for Reducing Stress

Go to the Bathroom One More Time

Most often, this initiative comes naturally. But it's better to include it in the planning as you may forget once you're focused on other parts of your preparation. On stage, it may grow into a distraction that you cannot do anything about! Additionally, seeking solitude in the bathroom provides an undisturbed moment to breathe and concentrate. Ensure your clothes are in order and don't forget to zip up your fly if you're wearing trousers. Take a look in the mirror, check your mouth and teeth for food stains. While you're at it, give yourself a big smile and assure yourself that you are one hell of a person and that the audience is privileged to have you![12]

Warning: if the organizers have given you a wireless microphone unit, ensure it is turned off before you go to the bathroom!

Manage Your Blood Sugar

During conferences and events, you rarely follow your normal eating habits. Irregular meals and a constant supply of buns and biscuits may lead to your blood sugar levels playing tricks on you, making you slow in the head and even dizzy. Therefore, ensure you eat according to a plan; slow-release carbohydrates (such as vegetables, legumes and nuts) are recommended.[13] During the talk, have a quick sugar fix at hand in case hypoglycemia hits, for example, soda (with sugar), sweets, or fruit.

---

[12] You may have heard about *power posing*, a technique where you allegedly should be able to increase your confidence and self-assertiveness by attaining a "superhero position." Unfortunately, the effect of power posing is a controversial issue; the authors of this book are skeptical, but we leave it to the reader to find their own opinion. Either way, striking a pose probably won't hurt!
[13] This is general layperson's advice. If you are a diabetic, follow the advice of your MD.

Remember, however, that this is just an emergency solution, as a large sugar intake will trigger insulin production, leading to a blood sugar rollercoaster.

Skip the Social Lunch
It can be hard to enjoy a meal before an important talk—you may feel locked in by social obligations, and the risk of swallowing a lot of air can cause bloating, stomachache, and a noisy belly. It may be better to eat alone in a quiet place—pick up a snack or bring your own food pack.

Look After Your Body
Your body is your main presentation tool—ensure it is fit for the job. Get enough sleep the night before your talk; don't drink too much alcohol, don't eat food that disagrees with you, allow yourself to rest. Dress comfortably: this does not mean wearing a tracksuit, but thinking about whether your jacket isn't too tight, or whether your skirt allows you to move freely.

Your muscles will likely be a bit tense before you go on stage. Loosen up by gently rolling your shoulders, and feeling the muscles around the top of your lungs relax. This little exercise will help release tension, and also give you something to focus your mind on.

Make Sure There Is Water
Performance anxiety (or sometimes even general excitement) is often associated with a sensation that saliva production diminishes and the mouth becomes dry. This is not only uncomfortable, but may lead to a feeling that it is hard to shape words and result in a hoarse voice. The solution is simple: stay hydrated (drink a large glass of water a couple of hours before your talk) and have water available during the performance.

Of course, you can have some other kind of drink, like soda or tea. But water is refreshing, doesn't stain if spilled, and doesn't promote tooth decay. A jug of water with ice cubes can also look really stylish on the podium—and can be used for playful improvisation.

Remember that the Audience Wants You to Succeed
For those who dread giving a talk, it may seem that there is a general hostility among the audience. For obvious reasons, this is not true: professionals attending science talks value their time and want to see interesting presentations. They don't want you to dwell on mistakes—they want you to proceed and keep the show running.

7.6.2 Handling Disturbances

One of the most stressing—and uplifting!—aspects of public speaking is that unexpected things keep happening. For example, one of us (Olle) once had his lecture interrupted by a window cleaner appearing behind him on a platform suspended outside the building. It is impossible to have a specific plan for such odd incidents. Instead, you need to develop a personal policy, a mindset, a set of rules—or whatever you want to call it—to help you improvise in a rational manner.

Here are some general attitudes that might help you diffuse a disturbance:

Approach it with humility and humor
If there is an evolutionary purpose of humor, it is to help us sweep aside emotions that might harm our social interactions. A shared laugh can defuse the most threatening or embarrassing situations, shifting the entire group into a different state of mind.

Let the show go on
Experienced musicians know one thing: if they play a wrong note, very few will notice. But if they dwell on their mistake, they will just increase the risk of another error. Again, we quote the classic British WWII poster: "Keep calm and carry on."

Maintain your dignity
If there is tension or conflict in the room, it is much easier to resolve the situation and make a successful exit if you stay dignified and don't let frustration and emotion get the best of you. Remain diplomatic, transparent, and sincere, and slow things down. If you don't know what to say straight away, it's perfectly acceptable to buy some time by excusing yourself for a moment by saying, "Sorry, it's a complex situation—I just need to collect my thoughts…"

If possible: deal with it later
If there is a hiccup, here and now might not necessarily be the best time to address it. To give the audience an uninterrupted experience, suggest solving the problem—whatever it may be—during the next break or at the end of the day.

Sometimes you may encounter individuals who are conflict seekers, know-it-alls, or even hostile. A good introductory line when someone is complaining is "It concerns me that …" or "I am sad to hear that …" such as:

- "It concerns me very much that you find my lecture pointless."
- "I am sad to hear that you are critical of the way I express myself."

The clever thing about this approach is that you promptly acknowledge the feelings or opinions of the dissatisfied, without admitting fault. This non-confrontational reply may actually be enough to remove the tension in the room. Often, people just want to have their feelings acknowledged, or demonstrate that they have a diverting opinion. If not, suggest a one-on-one discussion outside the venue room; based on our experience, this is usually accepted.

In very rare situations, a dissatisfied person decides to go all in and refuses to let you continue. The most drastic action you can take is to turn the rest of the audience against them by asking, for example, "What do the rest of you think—should we change the schedule and discuss this matter?" Most likely, the answer will be "No!" But proceed with caution: this solution could blow up in your face if the rest of the group agrees with the complainer, putting you in an uncomfortable situation where you might need to involve a colleague or manager to resolve the issue.

To sum up: stay on track, keep calm, remain polite, believe in yourself, and trust your sense of humor.

7.6.3 Some Useful Precautions

To avoid interruptions happening at all, you can prepare yourself in several ways. You've probably seen your fair share of presenters struggling with incompatibility issues (think: cables, connections and formats).

Never Trust the Technology in the Venue
Make sure you have a backup of your presentation online and/or on a memory stick, in a format that can be shown in any web browser or PDF viewer. Additionally, prepare for a situation where you cannot connect to the local WiFi. Learn how to connect your laptop to the internet via your mobile.

Bring Your Own Hardware
If you frequently give presentations, it is a wise thing to carry two parallel sets of critical equipment: dongles (if needed) and power adapters (either two of them, or one power adapter and one big powerbank). Don't store everything in the same place—you can, for example, keep one set in your suitcase and one in your rucksack. Bring your own, trusted remote. Finally, an extra HDMI cable, a USB memory stick, and other essentials can be the basis for an improvised MacGyver solution.

Learn How to Do Without
When you have some spare time, plan how you would give your lecture, seminar, or workshop without the aid of PowerPoint slides or a working internet connection.

Acknowledgements Thank you to Lydia Flock for contributing to this chapter.

References

Milson, J. (2016, August 10). *"Judge softly" or "walk a mile in his Moccasins"*—by Mary T. Lathrap. James Milson—Writing & Things. Retrieved March 4, 2020, from https://jamesmilson.com/about-the-blog/judge-softly-or-walk-a-mile-in-his-moccasins-by-mary-t-lathrap/

Morgan, N. (2014). *Give your speech, change the world: How to move your audience to action.* Harvard Business Review Press.

Quiller-Couch, A. T. (1916). *On the art of writing.* GP Putnam's Sons.

References

Rosling, H. (2010, December). *The magic washing machine.* TED. Retrieved April 10, 2020, from https://www.ted.com/talks/hans_rosling_the_magic_washing_machine

Flock, L. (2021). *Oxford vocal massage.* Oxford Vocal Massage. Retrieved April 15, 2020, from https://www.oxfordvocalmassage.co.uk/

Wikipedia Contributors. (2019, April 22). *Devil's advocate.* Wikipedia; Wikimedia Foundation. Retrieved April 20, 2020, from https://en.wikipedia.org/wiki/Devil%27s_advocate

Producing Quality Writing for a Range of Purposes

8

> **What you will learn from this chapter**
> Becoming an effective and reliable writer requires effort, focus, and perseverance. Not only do you have to master the technicalities of, say, structure, clarity, style, or tone of voice, but you also have to harness your creativity to inspire, surprise, and motivate readers.
>
> In this chapter, you will
>
> - learn about the essence of good writing,
> - receive some practical tips on how to get started or what to do when dealing with writer's block,
> - become familiar with writing techniques such as chunking or the inverted pyramid.

8.1 A Note to the Human Writer, Considering the Robot Writer

As we noted in Chap. 2, the tsunami of generative AI tools that followed the launch of ChatGPT has become a huge game-changer for many communication avenues. This disruption has, unsurprisingly, spread some confusion among the production team of this book. To handle the situation we have decided the following:

- Sections 8.2–8.5 are written from a pre-ChatGPT perspective.
- Section 8.6 reflects on how things have changed due to the advent of gen-AI and offers some hands-on advice on how to use it in a way that is not only practical and effective, but also ethical and dignified.

8.2 Skilled Writers Have an Amazing Tool of Persuasion at Their Hands

If you're not Swedish, you've probably never heard of August Strindberg (1849–1912). Yet this Stockholm-born author and playwright is considered one of the finest writers of nineteenth-century Scandinavian literature. Strindberg ruthlessly exposed the vices and hypocrisy of the bourgeoisie, and continuously stung the rulers with his sharp pen; he was in permanent opposition to—well—everyone! How Strindberg perceived his own role is hinted at in the poem *Lokes smädelser*,[1] where he lets Loki—the trickster of the Norse gods—declare: *You may be in charge—but I have the language, yes, I am in charge of the language!*[2]

Being in charge of the language is truly a superpower, worthy of gods. Skilled writers who master composition, form, clarity, rhythm and style, have an incredible tool of persuasion at their fingertips. Just think about it: the vocabulary of any human tongue is a treasure trove[3] from which countless word combinations can be formed to engage readers in innumerable ways, for example assuring, affecting, advising, affirming, alluring, or animating them. Grammar should not be seen as something limiting; instead, it is a rule book of possibilities offering countless ways of fine-tuning sentence syntax to make meaning and purpose either crystal clear or deliberately ambiguous to suit the communication task.

You might ask: "Since this is an English text, why not simply use the expression 'The pen is mightier than the sword' to make your point?" Well, firstly: as that is a very common expression—almost a cliché—it has lost much of its glow. In fact, it could be considered a "dead metaphor," which, according to Wikipedia, is "a figure of speech that has lost the original imagery of its meaning by extensive, repetitive, and popular usage."[4]

Secondly: as you may have noticed, we constantly use the principles of salience and stickiness (see Sect. 5.4) in this book to make our point: don't let your audience fall into a lull where they encounter no surprises, or any changes in pace or perspective. We could call this the *Beatles principle*, as every track Paul, John, George, and Ringo produced after 1964 had an interesting riff, harmony, lyric, or arrangement. In this way, every single Beatles song from their later career has a distinct personality. Be a smart writer: be like The Beatles!

---

[1] '*Loki's insults*'.

[2] "*I han makten, jag har ordet, jag har ordet i min makt.*" Verbatim, it's "You have the power/I have the word/I have the word in my power."

[3] The word *thesaurus*, that is a reference book for finding synonyms, actually comes from the Greek word *thēsaurós* which means 'storehouse; treasury'.

[4] Wikipedia Contributors (2017).

> **Written communication tips in this book**
> Besides the general overview here in Chap. 8, practical instructions and advice for writing tasks can be found in both Part II and Part III of the book.
>
> - Section 6.4: Telling a Good Story
> - Section 11.2: Articles for Peer-Reviewed, Scientific Journals
> - Section 11.3: Writing Effective (One-on-One) Emails
> - Section 11.4: Grant Proposals
> - Section 11.5: Press Releases
> - Section 11.6: Writing for Mainstream Media
> - Section 11.7: Writing an Interview for a Magazine, Newsletter or Blog.

8.2.1 Defining Good Writing as Effective Writing

So, what is good writing all about? The best definition we've read so far is found in a web writing guide by the British editorial trainer and consultant Susannah Ross.[5] Her statement aligns perfectly with the "doing-things-with-words" approach of this handbook:

> *Good writing is writing that does its job.*

Ross explains that "[Good writing] achieves its purpose, whether that is to tell you how to operate a lawn mower, to persuade you to book a holiday or to move you to tears over the suffering of a person you don't even know."

This reflection points back to Chap. 3 where we underlined that every communication activity should have a goal. Defining this goal during the writing process instantly takes you much closer to the finished text. Below, we list some examples of what the "job done" actually means for different types of texts.

Scientific journal article
"Job": to present a research group's proceedings to the scientific community in a way where the research question is well-defined, the experimental procedure is replicable, and the analysis and conclusions are transparent.

News article (in serious news outlets)
"Job": to deliver clear, well-structured, concise, objective, and useful information on recent events that are relevant for the reader.

News article (in tabloids)
"Job": to stir strong emotions, for instance anger, frustration or anxiety, to sustain the revenue-generating advertising mechanisms of the medium.

---

[5] Ross (2022).

Poem
"Job": to create feelings, and explore the cosmos, culture, and human condition in a way that is full of connotations. Unambiguity is not desirable.
Manual
"Job": to ensure a user can operate and use the features of their newly purchased appliance, system or service. Unambiguity is highly desirable.

Professional texts come in a number of formats, representing different purposes and communication channels. For each new format you learn to master, you will consolidate and develop those you have already mastered. See Chap. 11 for a selection of recipes for different writing scenarios.

8.3 The Craft of Writing

When it comes to pole vaulting or playing Paganini on the violin, there is a strong association between ability and virtuosity: those who manage to perform these feats are basically the same people who can be considered skilled in their art. Writing, however, is different. Most people in the industrialized world know how to write, and they produce a number of texts every day—from snippets in social media to emails, memos, and blog posts. Still, it is takes much more effort, focus and perseverance to become a writer who stands out from the crowd—a pro who is in control of structure, clarity, style, and tone of voice, and who inspires, surprises and motivates the reader with the right level of creativity for the task at hand.

So, how do you become this kind of writer? Well, at the core, you should engage in three activities:

- Activity 1: Read a lot—all the time, and all sorts of genres. Be bold, kill your prejudice, and stretch beyond your comfort zone.
- Activity 2: Write a lot—all the time, and in all sorts of genres.
- Activity 3: Reflect, discuss, and listen to the voice of experience—all the time and across all genres.

We cannot help you further with Activities 1 and 2 above, beyond wishing you good luck, but we can offer some advice for Activity 3. We have organized our initial suggestions and reflections on two levels:

- **8.3.1 The essence of good writing**
  The characteristics of a well-crafted text—that is, the outcome you should strive for.
- **8.3.2 Making your writing do its job**
  Guidelines you should have in mind as an efficient writer.

Advice from the best

All over the world, renowned writers and linguists of different mother tongues have engaged in the art of writing and the development of their languages. In some countries, it is the highest honor to become a member of a language academy, that is, a body that regulates a language. The most famous example—and The Enlightenment Age inspiration for them all—is L'Académie française, founded in 1634. But there are plenty of them all over the world: Svenska Akademien (Swedish), 국립국어원 (Korean), Rada Języka Polskiego (Polish), Asociación de Academias de la Lengua Española (Spanish), and so on. Surprisingly enough, an official body like this is lacking for the English language.

Many prominent authors have shared their personal reflections on the craft of writing. Some popular titles in English are *Zen in the Art* of Writing by Ray Bradbury,[6] *On Writing* by Stephen King,[7] and *On writers and writing* by Margaret Atwood.[8] Shorter pieces which you can look up are, for instance:

- Jack Kerouac: *30 Beliefs and Techniques*.[9]
- Henry Miller: *11 Commandments*.[10]
- David Ogilvy (copywriter): *10 No-Bullshit Tips*.[11]
- George Orwell: *Six Rules for Writing*.[12]
- John Steinbeck: *Six Pointers*.[13]
- Kurt Vonnegut: *Eight Tenets of Storytelling*.[14]

8.3.1 The Essence of Good Writing

How do you spot a well-crafted text? The simplest answer is this: it's when the text is a joy to read—no matter how interested you actually are in the subject. If we were to point out four characteristics of such excellent writing, they would be:

Characteristic #1: Well-structured paragraphs

A good writer knows how to put words together.

A great writer knows how to build sentences.

But an excellent writer knows how to form paragraphs!

---

[6] Bradbury (2017).
[7] King (2000).
[8] Atwood (2015).
[9] Kerouac and Weinreich (2009).
[10] Miller (1964).
[11] Ogilvy and Raphaelson (2012).
[12] Orwell (2013).
[13] Steinbeck (1975).
[14] Vonnegut (2000).

A well-formed paragraph starts with a *topic sentence* that makes a statement—like this one just did. The statement is then explained and elaborated upon. Once the writer is sure that the reader has grasped the point, the next step could be to add some "nice to know" information. In this case, we could, for example, mention that the Swedish expression for topic sentence is *ankarmening*, which means 'anchor sentence'. Finally, an elegant touch is to add a sentence at the end that points to the next paragraph; this gives a forward momentum for the reader. Consequently, you may notice that the next paragraph also starts with a topic sentence.

An intriguing thing about topic sentences is that if you line them up on a page, they form a summary of the main content of your text. In fact, the topic sentences could serve as your speaker's notes if you were to present the entire text to an audience—or as an abstract for your scientific article. If we turn this principle upside down, it is not hard to see that a convenient way to start planning a text is to set up a sequence of topic sentences, even a hierarchical structure (see Fig. 8.1). A lucid analogy would be the table of contents of a doctoral dissertation. This approach can be applied, for example, when writing a report of some kind (read more in Sect. 11.2).

From mindmap to text

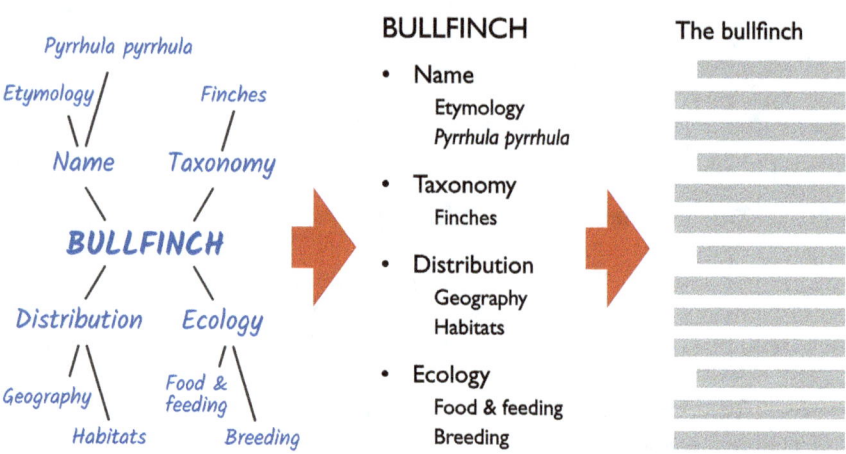

Fig. 8.1 A mind map may be the first step to write a well-structured text with well-formed paragraphs, starting with informative topic sentences

Characteristic #2: Rhythm and pace

Are short sentences better, or are long sentences better? The answer is: neither. Instead, sentences should have rhythm, a little swing. On the one hand, you can use the kind of long, meandering sentence you are reading right now to express something that is interesting in its complexity, and, as you note, the pace seems to increase as it goes on and on, but is still easy to read as everything is lined up in a sequential way that doesn't put a heavy load on your short-term memory; adding a second part after a semicolon helps us make the sentence even a little longer without losing you, dear reader. On the other hand, you can change pace, just like that! Slow it down. Stop to think. Keep still. Silence. Now, as we are sure that you have gotten the point, let's see what the next paragraph has in store for us.

Characteristic #3: The right amount of variation and surprise

Excellent writing is just like a good presentation—or a good piece of music, for that matter: it has variation and dynamics when it comes to rhythm, style, and other dimensions. It also keeps the reader awake and curious by coming up with unexpected things (like the August Strindberg quotation at the start of this chapter). The level of creativity should also vary: excellent writing is inventive when it needs to make a point, but uses standard language when that serves the purpose of the text better. For example, a popular history writer may choose a documentary, action-filled style while describing dramatic events, but a very plain and well-structured style for overviews and background explanations.

Characteristic #4: One-liners

Good writing often contains single sentences or paragraphs which stand on their own and may be quoted without the need for further context. From this perspective, your most essential writing tool may not be the word processor, but the pen and the notebook you carry in your pocket. When inspiration hits, when insight strikes, or when something extraordinary happens before your eyes, make sure you take notes. Inventing one-liners is a bit like playing Pokémon Go: you never know when an *Articuno* or *Lapras* will show up during your walks.[15]

8.3.2 Making Your Writing Do Its Job

What are the best strategies for putting words on a page? The answer to that question is the annoying "it depends." And it doesn't depend on only one factor, but on a number of them: the purpose of your text, the context in which it will be published, the genre, the available word count, and—not least—your personality. Nevertheless, we'll try to come up with some universal advice.

There is no single good way to produce a text.
We're all different, therefore there isn't a single method that suits everyone.

- Some people prefer to write a little every day, while others work in bursts.
- Some people start at the beginning and write straight through to the end, while others write a little here and a little there, and glue the pieces together during the review process.

---

[15] According to our Pokémon Go consultants, the species mentioned are rare and elusive.

- Some people write a plan or outline and stick to it, while others start somewhere where it is easy to start, and build up the writing project as they go.
- Some people work best in the early morning, whereas others prefer to burn the midnight oil.

Finally, a defense of the procrastinating mind. It is generally considered a sign of a careless and sloppy character to delay work to the last evening before the deadline, or to multitask. But what if this is the best way to produce quality output for the individuals in question? There is no reason to turn work methods into a discussion about personal virtue; if it ain't broke, don't fix it!

Feedback increases speed, accuracy and general quality.
Right away, we must contradict ourselves[16] (see above), and claim that there is actually one method that improves the writing process for all, namely this: early and frequent feedback. This principle has been thoroughly examined via the so-called Agile methodology—a development method widely used in the software industry. If the software team continually confirms that they are following the right course, they will work faster and produce better results.

This means that you should engage your supervisor, your mentor, your editor, or your friends as early and as much as practically possible.

> **The art of giving feedback**
>
> **Match the ambition level with the project status.**
>
> Always adapt feedback to where the writer is in the production timeline. Don't correct typos in an early draft, and avoid meddling with the organization of the text in a late draft.
>
> - Early feedback: Are the instructions of the given assignment generally followed? Is the content organized in the most effective manner? Is the writer sticking to the conventions and rules (formal or informal) of the genre? Is the level or style appropriate for this type of text?

---

[16] "Do I contradict myself?

Very well then I contradict myself,

(I am large, I contain multitudes.)"

Walt Whitman 1819–1892: Song of Myself, 51 (Whitman, 2005).

- Intermediate feedback: Have the instructions of the given assignment been followed in detail? Include: (a) more detailed feedback regarding the content and the organization; (b) stylistic feedback, suggesting improvements in clarity and style; (c) corrective feedback, pointing out mistakes in grammar and spelling.
- Final feedback: Has the writer finished the assignment, or delivered a text ready for publication? Now is the time for word-by-word examination to find any remaining mistakes in grammar and spelling, as well as typographical errors (professionally referred to as proofreading).

Be specific.

Give instructions and suggestions that are easy to follow and understand. Avoid comments like "A bit vague here" or "How about elaborating on this?" Instead, suggest concrete ways to make the text clearer, or specify what information should be added.

If you like it, don't forget to say so!

For the person on the receiving end of feedback, a long list of comments can be dispiriting, forecasting hours of work on the more tedious aspects of writing. Worse still, comprehensive feedback—with the best of intentions!—can even turn out disheartening, creating self-doubt and disappointment. To avoid this, don't forget to give encouragement and praise—both on a general and specific level. The burning questions many writers don't dare to ask are: "Besides your comments about this and that, did you LIKE it? Was it any GOOD? Am I USELESS, OUTSTANDING, or perhaps something in between?"

There are thousands of ways to solve every writing task in a way that is good enough, great, or sometimes even brilliant.

The number of permutations in which you can combine all the words of your language—dictionaries, professional terminology, colloquialisms, etc.—in a grammatically correct way is astronomical. Or perhaps we should use adjectives like "monstrous," "mountainous," or "mammoth," as this fact is somewhat intimidating, even a bit scary.

Every word and expression we add to our document while writing could be replaced by another ... no, exchanged for an alternative ... no, picked differently ... If we're haunted by an idea of achieving perfection, this immense freedom of choice may seem both stressful and frustrating.

Here is some good news: there is no single text which is superior to all other possible versions. However, there are plenty of versions that "do the job," according to the Ross principle above. Many of them are even great, not to say brilliant. But none of them are reliefs on the wall of Plato's cave, representing the idea of the quintessential text.

The French poet Paul Valéry once stated that a work of art is never truly completed, only abandoned. And the American novelist Herman Melville included a kind of a meta-comment in his 1851 book *Moby Dick*: "God keep me from ever completing anything. This whole book is but a draught—nay, but the draught of a draught."[17] Despite his reservations, we all know the novel in question turned out sort of OK in the end.

Superfluous detail obscures your main point.
A zealous attention to detail is both a blessing and a curse among people of science. It is a great trait when setting up and performing experiments, analyzing data, finding correlations, and generally squeezing elusive reality into a scientific model. However, its value is ambiguous when it comes to communication. In our experience, many scientists—in all fields—tend to hide the shimmer and beauty of their findings behind a trite fence of humdrum detail; this is true for both their oral presentations and their writing. Journalists who help scientists popularize their research often have to put in a lot of effort to make their clients let go of the nitty-gritty and instead focus on the main findings, connections, and consequences.

8.4 Approaching a Text Assignment

Getting started is often the hardest part. Below are four ways to approach a text assignment, followed by a few tips on combating that horrible feeling of being stuck.

- 8.4.1 Starting with the summary
- 8.4.2 Brainstorming + Outline
- 8.4.3 Orientation, Information and Action
- 8.4.4 Drawing an article
- 8.4.5 Managing writer's block.

8.4.1 Starting with the Summary

What do you want the reader to remember from your text? Just as when you prepare a presentation, things will probably fall into place once you know *what* you want to say, as well as *why* you want to say it. The principle of presenting the essence of your text in just a few sentences is sometimes called "nutshelling." To achieve this, start by doing two things:

---

[17] Melville et al. (2021).

- Clarify your objectives. What is the purpose of the text—are you out to inform, debate, persuade, entertain …?
- Define your main messages (for example, "I want to convey the following three standpoints: …"). Alternatively, identify the main content (for example, "who did what at which point in time…").

Similarly, if you're writing a scientific article, start by drafting a quick version of the abstract—what are your key questions, and do your key findings match them? Or, if you're writing a journalistic article, start by putting together a lead paragraph that seems to work well. For a report, start with a sketch version of the executive summary. Helpful methods include the Five Ws in Sect. 4.4.3 or the elevator pitch protocol in Sect. 10.2. Alternatively, you can start by filling in the following sentence: "What I want to show the world is …"

Here's an anecdote which is interesting in this context. Some years ago, Olle spoke with a scientific writing coach who used to proceed in the following way during her first session with clients (typically scientists who wanted to improve their writing, using their latest research project as material). As they sat down to begin their discussion, she pushed the printout of the text draft gently to one side, and instead asked: "Please tell me: what are the three takeaway messages from this study?" The scientist thought for a while, and came up with three conclusions. "OK," she said, "let's now have a look at your abstract." As they were reading, the scientist interrupted and exclaimed: "Wow, I see now that I have not included these takeaway messages!"

8.4.2 Brainstorming + Outline

This method is similar to the sticky notes method used to prepare a presentation in Sect. 7.2.

1. **Set up a scope**
   What is the goal of the text? What are its main messages or core content?
2. **Brainstorm**
   Choose a room where you feel at ease and no one disturbs you, grab a deck of sticky notes and a nice pen/pencil, and start writing one item on each note. Try to approach the subject from different angles. Step into the shoes of the reader: how much do they know about the subject, what makes them curious, and what do you want to convey to them?
   Once you start running out of ideas, challenge yourself, for example by deciding to add ten more notes, or three more notes about the methods. When there is no more low-hanging fruit, you need to bring out the ladder of creativity.
   If you are the type of person who is constantly jotting down ideas, bring your notebook to the brainstorming session and transfer those ideas to your growing sticky note collection.

3. **Add structure**
   You have concluded the brainstorm; now it is time to get organized. Start by clustering the sticky notes, then add stickies that describe each cluster as a category.
4. **Create a written outline**
   Transfer the sticky note mind map into a hierarchical list on your computer. Feel free to integrate this process with the idea of topic sentences, as described in Sect. 8.3.1 above.

8.4.3 Orientation, Information and Action

Orientation, Information and Action (O-I-A) is a concept first introduced by writing coach Crawford Kilian in his book *Writing for the Web*.[18] Building on the idea of a well-formed business letter, Kilian stated that a web text should provide an appropriate background (Orientation), help the reader perceive the main message (Information), and understand what should happen as a result of the information conveyed (Action).

This structure is excellent for emails, web pages, invitations, promotional posters, etc. Read more on writing effective emails in Sect. 11.3.

8.4.4 Draw an Article

If you're the kind of person who is especially fond of visual explanations and planning, you could use pencil, paper, and sticky notes to create a one-pager of your article (see Fig. 8.2). This may also come in handy if your publishers ask for a graphical abstract or a social media visual after acceptance.

---

[18] Kilian (2015).

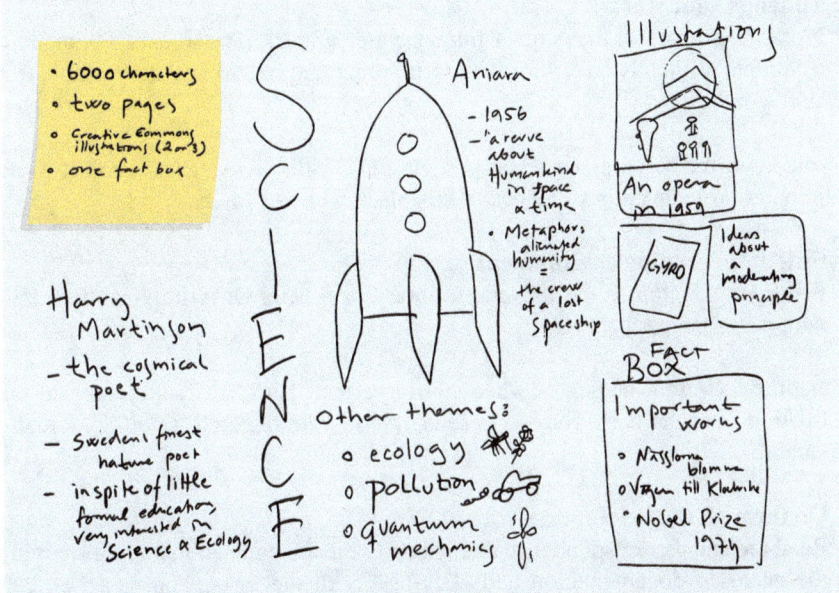

Fig. 8.2 Olle's one-pager, summarizing the life and works of Harry Martinson—a Swedish poet who often explored scientific, ecological, and technical themes

8.4.5 Managing Writer's Block

We've all experienced the tyranny of the blank page. Sitting and staring can be just as tedious and demoralizing as constant writing and deleting. This situation should be taken seriously, as writer's block can not only mess up our weekly work life, but also inhibit our professional progress and, in severe cases, affect our self-esteem.

Rule number one for managing writer's block: don't stay stuck! When you've made an honest attempt to get started and failed, leave your chair and do something else. The longer you sit and feel like a failure, fantasizing about the consequences of your inadequacy, the harder it will be to escape the situation.

Loosen up
At this point, the stressful situation has triggered a response similar to performance anxiety—see Sect. 7.6 for managing your nerves. You need to reverse this process by either relaxation (mindfulness, yoga, meditation) or physical activity (a brisk walk or a run, a 7-min workout).

Sharpen up
Your inability to focus may not only be stress-related. Perhaps you need a snack (preferably slow carbohydrates) or a power nap.

Timebox your work
Work in sprints: full focus for X minutes, then a break. The so-called pomodoro technique originally called for 25 + 5 min, but feel free to find the formula that suits you best.

If you sit in the same place when you struggle with reluctant text projects, you will condition your brain to associate this place with negativity.

Switch the writing environment
Bring your laptop to the university library or a café. Or write your first draft with pencil and paper.

Sometimes, we fool ourselves when starting a text. Have we really received the instructions and the background we need? And is the research we've done really adequate?

Go through the brief
Read the email correspondence that led up to the assignment, including studying the enclosed documentation and clicking on all the links. You can even call the editor or whoever is the next person in the editorial chain for a clarifying conversation.

Sit down to read
Your slow progress may be due to not fully understanding or internalizing the subject. Make some printouts, fetch books from the library (or just grab your laptop) and sit down in a comfy environment to read. Bring highlighters and sticky notes!

Our experience shows that self-esteem is extremely important for the writing process. That's because writing is about making hundreds of small decisions every hour, and if you keep hesitating because you don't trust your skills or judgment, your progress will be slow.

Do something that makes you feel like a boss
Play a (short) level in a video game in which you normally excel. Call a person who likes you, and have a fun conversation. Read some of your stuff that you feel proud of.

Write for someone who believes in you
A well-known tip for nervous public speakers is to focus on a friendly face in the crowd. In a similar way, write for someone you know will like your text.

Alternate with some pleasure writing
Work on two texts simultaneously: your problem text and one that is fun to write.

8.5 Some Tricks of the Trade

We have been through some useful principles already, such as the rule of three (Sect. 4.4.4), the principle of salience (Sect. 5.4.2), and storytelling techniques (Sect. 6.4). Here, we add a few more ways of working that may be helpful for you—right after a quick reflection on the English language.

> **English grammar, punctuation, and style**
> "The wonderful thing about standards is that there are so many of them to choose from," quipped an ironic person[19] once upon a time. The same statement applies to the rules and guidelines of the English language.
>
> Of course, it is no easy task to define, order and apply coherent grammar and style rules for a language with 600,000 thousand words, spoken officially in 67 countries. The task doesn't become easier when 400 million native speakers and almost 800 million second-language speakers are joined by 700 million people who use it as a foreign language. Consequently, there is no obvious homestead for this language; instead, local varieties are developing all over the real and the virtual world.
>
> In short, here's what you need to be in control of:
>
> - **Grammar, syntax, and punctuation**: standards and conventions of what is considered right or wrong when constructing sentences in English.
> - **Spelling**: same as above, but at the word level.
> - **Style**: what sounds proper, idiomatic, and current in English; this can be elusive and hard to master if you don't live in an English-speaking environment.
>
> **Be consistent**
>
> Choose one type of spelling (American, British …) and stick to it. Similarly, take your pick among the grammar and style guides (Chicago, APA…) and dictionaries (Merriam-Webster, Oxford, Collins …). During your career you will notice that different organizations may have their own guidelines, so always ask for them; there might even be a local style guide.
>
> **Learn punctuation thoroughly**
>
> Knowing all the rules of punctuation helps you in two ways. Firstly, it turns you into a better writer by improving your ability to direct the reading experience. Secondly, it makes the editorial process much smoother: there's no need to change what's already correct.

---

[19] This person is either U.S. Naval officer and an early computer programmer Grace Hopper, or computer scientist Andrew S. Tanenbaum. Or perhaps someone else?

> Even if your native language uses the Latin alphabet, its punctuation rules may differ from English. Let's use quotation marks as an example:
> English: "quotation"
> Swedish: "quotation"
> German: „quotation"
>
> **Be careful with diacritics**
>
> English uses very few diacritics (the little marks placed above or below letters to indicate a special pronunciation). Authors used to writing exclusively in English thus tend to see them as insignificant. However, in some languages, it is more problematic to omit them than in others. For example, a Polish reader wouldn't have any problems reading *czesc* as the correct *cześć* ('hello'), but in Swedish, *hal* means something totally different than *hål* or *häl* (*'slippery'*, *'hole'*, *'heel'*).
>
> **Use synonyms**
>
> As soon as you feel that the linguistic landscape on your page is bogged down by uninspired wording, stereotypes, and repetitions, go to the fresh well of your thesaurus to revitalize your writing.

8.5.1 Chunking—A Cornerstone of Modern Writing

From personal experience, one of us (Olle) has formed the opinion that there is no quicker way to learn and memorize history than by browsing through popular history magazines of the serious but fun type—for example BBC History Magazine (UK), True West (US), or Populär Historia (SE). These magazines manage to convey so much complex information in a manner that is effortless for the reader. They use modern feature writing style and storytelling, and they rely on visual communication with photos, illustrations, maps, and timelines. Most importantly though, they are experts in chunking.

Wikipedia defines chunking as "a method of presenting information which splits concepts into small pieces or 'chunks' of information to make reading and understanding faster and easier."[20] We know from cognitive psychology that memorization is enhanced when related items are grouped into manageable chunks. This works on all scales—for instance, memorizing a phone number as 22-33-6060 is easier than 2-2236-060, as the first instance follows a more obvious pattern.

How would we use this tool in writing?

Firstly, think about your paragraphs: are they short, snappy and clear? Each paragraph should focus on a single idea or topic. If the format permits, use subheadings to signpost the content. These natural chunks guide readers through the material.

---

[20] Wikipedia Contributors (2019b).

Secondly, while we mentioned that varying sentence structure is key to maintaining a nice flow, try to avoid overly convoluted sentences. Smaller, more coherent units are more effective, especially when conveying complex subject matter.

Thirdly, separate chunks visually. This could mean either breaking a wall of text with a bullet point list or a subheading, or adding a visual—graph, figure, or text box—to draw the reader's attention to a single chunk of information.

What to do with too much text?
If the text you have just written seems useful, but goes beyond a word count, don't panic—there are still ways to incorporate it.

Is it interesting for context but hard to include in the main text?

→ Write a fact box.

Is it related to a photograph or an illustration?

→ Move it from the main text into a long caption.

Does it reflect causality or progress?

→ Add text in a graph or timeline.

Has someone else done a better job with a succinct description?

→ Include a quotation.

Is it great stuff, but not quite to the point?

→ Move it to a "darling bin"—a folder for storing all the wonderful work you produced that didn't quite make the cut (but can nonetheless be used later, for other purposes).

8.5.2 "The Inverted Pyramid"

What is the single most effective skill you can learn to become a more effective communicator both within and outside science? One of the top candidates must be mastering the Inverted Pyramid format—the basic principle of writing brief news articles.[21] To explain why, let's consider an example:

> *Imagine meeting a colleague by the coffee machine at work; both of you are in a bit of a hurry. Your colleague asks: "How was the conference in Boston last week?" What would your reply be?*

---

[21] Roberts (2016).

> Well, probably not this: *"John and I started the journey last Sunday by getting up at five o'clock and taking his Toyota to the airport. There, we had to wait for three hours or so. To pass the time, we ate breakfast. The flight took about five hours to Reykjavik ..."*
>
> Instead, your friend is expecting something like this: *"It was a great experience. I learned a lot about epigenetics in alcoholism, and we had productive discussions with Dr. Lidenbrock and people from her group when we hung out in the hotel lobby two nights in a row. We're scheduling a call with them soon."*

In the example above, the first reply is, of course, useless. If your friend doesn't stay put and listen for ten minutes, she will not learn anything relevant. The second reply, on the other hand, nails some main messages right away: (1) the trip was worthwhile; its outcomes included (2) knowledge and inspiration in a specific field, (3) a deepened relationship with important peers, and (4) arrangements for next steps towards collaboration.

This "straight-to-the-point" reply demonstrates what news journalism is all about: knowing what the receiver wants, and delivering it in a concise way. Now imagine that the conversation would continue for a minute or two: what would the natural follow-up be? Your friend would probably want you to elaborate on the information you've already given to fill in the blanks. With each question, the conversation would go into more and more detail. Ta-da: you are now following the inverted pyramid format used in news journalism!

But let's take it from the beginning. The Inverted Pyramid is a principle followed by countless journalists, copywriters, PR writers, press officers, etc. Nielsen Norman Group define it like this:

> *In journalism, the inverted pyramid refers to a story structure where the most important information (or what might even be considered the conclusion) is presented first. The who, what, when, where and why appear at the start of a story, followed by supporting details and background information.*[22]

Since the nineteenth century, the inverted pyramid has become the de facto standard for writing news stories. Additionally, it is useful for press releases, blog posts, editorial content, data sheets, web catalogs, advertisements—yes, even for things like manuals and instructions, public information, and official announcements (Fig. 8.3).

---

[22] Schade (2018).

8.5 Some Tricks of the Trade

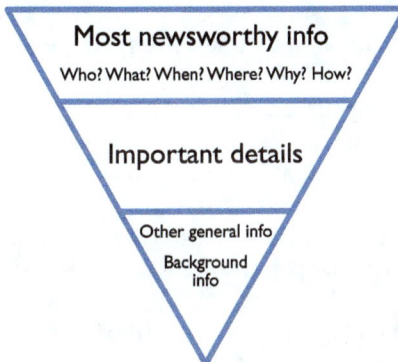

Fig. 8.3 The inverted pyramid has become a standard method in the media industry and is taught to all students of journalism, media, and communication. (Illustration adapted from Wikipedia[23])

The inverted pyramid format offers a number of benefits for writers, editors, and readers[24]:

- For writers: it speeds up the research, selection, and writing process.
- For editors: not only do they quickly grab the main point—they can also easily shorten pieces by simply deleting the last paragraphs.
- For readers: It helps to find key information quickly and navigate through content, allowing them to select how much detail they want to go into. When they have had enough of a subject, they simply skip the remaining paragraphs.
- For everyone: It creates a standardized presentation format familiar to writers, editors and readers, where relevant information is always the focus.

> **The inverted Pyramid—a part of media history**
> Before the advent of the telegraph in the 1850s, journalistic news texts were quite different from what we read today. Reporters routinely took on the role of travel writers or storytellers, providing numerous details to create atmosphere and spatial context, using the stylistic techniques and mannerisms of authors describing dramatic events in their books. For example, this is how William Howard Russell—the world's first war correspondent—began his report on the disastrous charge of the Light Brigade at Balaclava in 1854:
> *"If the exhibition of the most brilliant valor, of the excess of courage, and of a daring which would have reflected luster on the best days of chivalry can afford full consolation for the disaster of today, we can have no reason to*

---

[23] Wikipedia Contributors (2019a).
[24] Lidwell et al. (2010).

> *regret the melancholy loss which we sustained in a contest with a savage and barbarian enemy."*
>
> The invention of the telegraph, first implemented in 1844, gradually changed this, and the reason was simple: economy. Sending texts by telegraph from distant places created amazing opportunities and increased the speed of the whole news business—but it was also expensive. Consequently, press offices developed the kind of language we associate with news journalism today: straight-to-the-point, objective, neutral, fact-oriented, and with short and snappy sentences.
>
> In the third millennium, the Inverted Pyramid is more relevant than ever.

Bear in mind that while the Inverted Pyramid is great in some scenarios (such as news articles or blog posts), it's completely useless in others. For instance, there's no point in using it to tell a story where you already have a captive audience willing to accompany you on a metaphorical journey. They don't want a spoiler ahead of the game! If you were to tell the tale of, say, Little Red Riding Hood, in an inverted pyramid style, it would start like this: "Huntsman saves little girl and elderly woman who have been swallowed alive by a wolf." Your audience wouldn't need to listen further—they already got the gist of the story.

8.6   Generative AI in Science Communication

When the handheld calculator replaced the slide rule in the 1970s, there were concerns that students would lose essential mathematical skills. When the word processor replaced the typewriter a decade later, there were concerns that vital language skills would be replaced by autocorrect. The public availability of Generative AI, which passed the Turing test with flying colors, has raised similar concerns. So, how do these tools affect you, a communicating scientist? And what can we deduce about the use of GenAI in written communication today that will still be valid tomorrow?

8.6.1   Addressing Concerns Over the Use of GenAI

Generative AI,[25] or GenAI, uses computational power and memory management to apply machine learning on large amounts of data called LLMs, large language models, while also adding information on context. The result is a statistical model

---

[25] Note that, in contrast, the term General AI (or AGI, Artificial General Intelligence) refers to machines that can interpret, learn, and act autonomously without the help of humans. These are encountered in science fiction, but still not in the real world, and will not be the topic here.

8.6 Generative AI in Science Communication

capable of analyzing and producing language for various purposes. While similar AI concepts can generate voice, image and video, this section focuses on text.

Let us start with a tip written by the infamous ChatGPT:

> "Incorporate AI as a powerful ally, but always infuse your personal expertise and passion to maintain authenticity in science communication."

But can tools based on statistical models be used without sacrificing your own personal voice and expertise? What are the risks? We will address these concerns from the perspective of classical rhetoric.

1. **Ethos**
   Generative AI is a mimicking technology with no agency of its own. It does not produce entirely new content, but builds on the existing content in its underlying data. This is not necessarily a bad thing—after all, we do owe our own success at least in part to the contributions of those who came before us.[26] But there is still cause for some caution. When using GenAI to produce content, treat it as a tool, not a blueprint, to avoid losing your ethos. You are unique, while Generative AI is not.

2. **Pathos**
   You want your communication to resonate emotionally with your audience. Remember that all AI models are prone to bias, inherent in both the underlying data and the contextualizing process. The lack of transparency in these data and the models applied to them makes it hard to assess the type and magnitude of that bias. When using AI to assist with communication, be aware of these potential flaws, and consider how to best connect with your audiences.

3. **Logos**
   From a purely fact-based perspective, we can address three potential impacts of using AI to generate content.
   Firstly, since the results are built on underlying data, there are concerns on infringements: for any substantial amount of text, you may need to cross-check that it does not include citations that breach existing copyright.
   Secondly, generative AI will sometimes produce counter-factual content, simply because it is based on a statistical model. Although improved GenAI quality reduces the risk of such "hallucinations," ultimately the responsibility for fact-checking your content rests on you.
   Thirdly, there are security concerns, particularly around data integrity. Your organization will (for good reasons!) have an AI policy, so dig it up, read it and stick to it—and add a sound amount of skepticism before sharing any and all data with GenAI tools. On top of that, always review the AI policy of the outlet for which you are writing, as it may vary between publishers. Their statement

---

[26] In 1675, Sir Isaac Newton famously wrote in his letter to Robert Hooke: "If I have seen further [than others], it is by standing on the shoulders of giants."

should be available online, in editorial policies or author guidelines. It is also advisable to check the guidelines of the Committee on Publication Ethics,[27] as well as advice from local, national and international ethics committees.

8.6.2 Potential Benefits of Using GenAI

Even bearing the above issues in mind, don't lose faith and optimism in new technologies. After all, we have enjoyed the use of calculators, word processors and many other technical advances for decades. So, going back to ChatGPT's recommendation, how might you best "incorporate AI as a powerful ally"?

1. **Adapt and summarize content**
   With a bit of practice, you can use GenAI to enhance productivity. GenAI can quickly summarize complex information and make it more accessible; it can also help you convert your initial notes into readable and coherent text. If you are slightly over a word limit, ask AI to shorten the text.
   To adapt your content, use GenAI to explain scientific jargon in lay terms, or to translate your text not only into another language, but also another cultural context.
   Let GenAI help when it comes to language and format; this will allow you to focus your efforts on the main messages and the actions you want to inspire.
2. **Manage inbound and outbound consistency**
   GenAI is good with standardized formats, which explains its early success in areas like weather forecasts, legal documents and sports results. Employ this feature in your inbound communication, like peer reviews, where you can save time by developing a template that is suitable for your way of working. This can help identify inconsistencies in results and conclusions, and summarize your reviewer notes into a standard format.
   For your outbound communication, use GenAI to ensure consistent messaging. For instance, if you frequently update an FAQ section, GenAI can help maintain a standardized process and format. It can also ensure consistency across communication channels. Just remember, AI is a powerful tool, but you are still the persona visible in your outbound communication.

> **Exercise VIII.I—a bit of AI play**
> Select a scientific article that has already been published, either your own, or any other article available in the public domain. Use a GenAI tool of your choice to:

---

[27] Committee on Publication Ethics (2019).

- Summarize the main contents for a peer.
- Explain the contents to a secondary school student.
- Apply basic rhetoric to the article, outlining the main ethos, pathos and logos components.
- Write a news article in 500 words, applying the Inverted Pyramid principle.
- Write a 14-line sonnet in the voice of Shakespeare, conveying the contents in a poetic manner.
- Propose the contents and layout for a scientific poster on the subject, designed primarily to attract external funding.

Not all the results might be equally impressive. Would you save or waste time using GenAI for each case? Are there other GenAI tools that will do a better job with some of these tasks?

References

Atwood, M. (2015). *On writers and writing*. Hachette.
Bradbury, R. (2017). *Zen in the art of writing*. RosettaBooks.
Committee on Publication Ethics. (2019). *Promoting integrity in research and its publication*. COPE. Retrieved May 21, 2024, from https://publicationethics.org/
Kerouac, J., & Weinreich, R. (2009). *You're a genius all the time: Belief and technique for modern prose*. Chronicle Books.
Kilian, C. (2015). *Writing for the web*. Self-Counsel Press.
King, S. (2000). *On writing: A memoir of the craft*. Simon and Schuster.
Lidwell, W., Holden, K., & Butler, J. (2010). *Universal principles of design, revised and updated: 125 ways to enhance usability, influence perception, increase appeal, make better design decisions, and teach through design*. Rockport Pub.
Melville, H., Said, E. W., & Tanselle, G. T. (2021). *Moby Dick*. Reader's Library Classics.
Miller, H. (1964). *Henry Miller on writing* (Vol. 151). New Directions Publishing.
Ogilvy, D., & Raphaelson, J. (2012). *The unpublished David Ogilvy*. Profile.
Orwell, G. (2013). *Politics and the English language*. Penguin.
Roberts, J. (2016). *Writing for strategic communication industries*. Ohio State University.
Ross, S. (2022). *Susannah Ross editorial training & consultancy*. Susannah Ross. Retrieved May 7, 2020, from https://www.susannah-ross.co.uk/
Schade, A. (2018, February 11). *Inverted pyramid: Writing for comprehension*. Nielsen Norman Group. Retrieved May 20, 2019, from https://www.nngroup.com/articles/inverted-pyramid/
Steinbeck, J. (1975). The Art of Fiction No. 45 (G. Plimpton & F. Crowther, Interviewers) [Interview]. *The Paris Review*. Retrieved May 15, 2020, from https://www.theparisreview.org/interviews/4156/the-art-of-fiction-no-45-continued-john-steinbeck
Vonnegut, K. (2000). *Bagombo snuff box: Uncollected short fiction*. Random House.
Whitman, W. (2005). *Walt Whitman's leaves of grass*. Oxford University Press.
Wikipedia Contributors. (2017, December 4). *Dead metaphor*. Wikipedia; Wikimedia Foundation. Retrieved June 3, 2023, from https://en.wikipedia.org/wiki/Dead_metaphor

Wikipedia Contributors. (2019a, April 25). *Inverted pyramid (journalism)*. Wikipedia; Wikimedia Foundation. Retrieved May 21, 2020, from https://en.wikipedia.org/wiki/Inverted_pyramid_(journalism)

Wikipedia Contributors. (2019b, August 13). *Chunking (psychology)*. Wikipedia; Wikimedia Foundation. Retrieved May 10, 2023 from https://en.wikipedia.org/wiki/Chunking_(psychology)

Designing Effective Visuals 9

What you learn from this chapter
We live in a world where visuals are an essential requirement in communication.

In this chapter, you will learn

- how to use visuals strategically,
- the basic rules of graphic design,
- how to show data in a way that is clear, explanatory, and honest.

9.1 Visual Communication is Essential Communication

Boosted by digital tools and the internet, visual communication (including illustrations, photos, diagrams, graphs, infographics, and videos) has become indispensable. This shift has not only widened the professional market for visual designers and producers but also created totally new areas of expertise; as a result, many amateurs and early-stage professionals can demonstrate impressive skills within this niche.

Visual communication has always seemed very natural in science: many things are studied, collected, analyzed and discussed through visual information. Data handling software provides us with a myriad of pretty graphs, maps, and figures used for analysis, demonstration, and clarification (though unfortunately, in some cases, also for manipulation). Additionally, there's a lot of spontaneous doodling and visual explanation during the constant, informal communication in labs, offices, and conference rooms—in notebooks, on whiteboards, and (our favorite!) on the glass of fume hoods.

> **Reflection IX.I**
> Have a look at the quick sketches you draw to explain things to your colleagues. Take some photos and study a collection of them, side by side.
>
> Do you have a particular style? Are there any hidden gems that you can use for discussions, presentations, teaching, or publications? Is something "lost in translation" between your quick sketches and the visuals you prepare for your slides, posters, and articles, for example interesting details or a fun, playful style?

Science, of course, has kept up with the above developments. Not only are we seeing new genres of scientific visual communication, like graphical abstracts, infographics and interactive lecture graphs—it is apparent that science itself has become more colorful and spectacular when it comes to both raw and processed data.

To help you start thinking more visually, we've put together the following sections containing tips in the common areas where you'll need graphics:

- Section 9.3: The Fundamentals of Design in Science
- Section 9.4: Guidelines for Reluctant Visual Designers
- Section 9.5: Visual Hierarchy Helps Us Organize Information
- Section 9.6: Making Typography Work for You
- Section 9.7: Choosing and Managing Colors.

> **Visual communication tips in this book**
> Apart from the tips contained in this chapter, other instructions for particular visual communication scenarios can be found in Part III of the book.
>
> - Section 12.2: A Slide Deck for a Short Scientific Presentation
> - Section 12.3: Designing an Effective Scientific Poster
> - Section 12.4: Producing Video Material, Such as an Abstract

9.2 Scientists Who Think in Design

Before diving into graphic design, let's take a moment to consider what the concept of "design" means to a scientist. On one hand, you might say that the blueprint of a steam engine is a design, with a clear emphasis on functionality. On the other hand, you can argue that a highly impractical piece of high fashion clothing is also a design, albeit with a focus on aesthetics. This duality may seem confusing, but, fortunately, modern definitions integrate both perspectives.

For example, Swedish design company Agero defines the verb "to design" as follows:

> "Based on a purpose, to create something that serves someone (yourself, other people, society and the world), using one (or several) needs as the starting point."[1]

Through this definition, we can see that a universal dimension of communicative design is its ability to create an impact.

A helpful approach to understanding this impact comes from an early theorist of Western architecture, the Roman architect, engineer, and writer Vitruvius (first century BC). He stated that all building structures should adhere to a harmonious triad of *durability, utility, and beauty*.[2]

Design is thus the intermediary that conveys that your scientific ideas **matter**. Your work deserves the attention and care that gives it the best chance to make the right impact. Well-thought-out, good design is simple and subtle, condensing only the most important and valuable information. Your audience will gain much more through your intentional design choices. It'll be worth the time and effort—you'll see.

Just as with writing, we recommend an iterative design process (see Sect. 8.3.2) where you let your ideas cool for a bit before you come back to work on them again. If you aren't very confident—and the final product is a high-stake one—start early to allow yourself the luxury of consulting your (friendly but honest) colleagues and incorporating their feedback. And remember: never let yourself be inhibited by the imperfection of early versions.

> **The Aristotelian triad in design**
> Graphic design is a discipline that takes a lifetime to master, particularly because it operates on two levels. On a basic level, the content itself—consisting of text, photos, illustrations—should speak as clearly and coherently as possible. On a meta level, the layout itself conveys a message. Achieve this dual goal through the ethos, pathos and logos model.
>
> **Ethos**: the design should tell the reader and viewer about the sender and what this individual or organization stands for. More than anything, it should establish the sender's trustworthiness—conveyed by a tasteful, neat, and restrained design.
>
> **Pathos**: the design should reflect the emotional mood of the subject and the attitude of the sender. A train timetable that looks like a children's book is just as useless as a children's book that looks like a train timetable.
>
> **Logos**: the design should present information in a way that is not distorted, to minimize the risk of misinterpretation.

---

[1] Agero UX Team (2023).
[2] If you're an Aristotle fan, you could see a correspondence here: durability and ethos, beauty and pathos, utility and logos. We're not the first to note this.

> **Examples**:
>
> A governmental information brochure about health screening programs should communicate seriousness, responsibility, and dependability. Therefore, the layout should be clear, elegant and minimalistic.
>
> A fanzine for fans of a hard-core punk band should put across a sense of community among outsiders who share the radical values and the "Up yours!" attitude of their beloved musicians. Thus, the layout here is chaotic and unconventional, featuring hand-written texts and grainy photographs.

9.3 The Fundamentals of Design in Science

Let's bring together Vitruvius' time-tested wisdom with the street-smarts of a new kid on the block in the field of communication: *information design*. This area of study and practice merges design perspective with communication theory and cognitive psychology. Basically, it is about making information as accessible and as easy to understand as possible. Information design is responsible for effective data visualization, public signs, user manuals, and so forth.

You can go a long way by letting yourself be guided by the three models below.

9.3.1 The Appeal, Comprehension, Retention Model

In their book,[3] designers Jason Lankow, Josh Ritchie and Ross Crooks propose a 21st-century version of Vitruvius's standards. They define good information design as having (1) appeal, while promoting (2) comprehension and (3) retention (Table 9.1).[4]

---

[3] Lankow et al. (2012).

[4] This could be compared to the three goals of effective presentations, defined by Stephen M. Kosslyn (2007): connect with your audience; direct and hold attention; promote understanding and memory. Note that appeal corresponds to the first two, whereas comprehension and retention are included in the last one.

9.3 The Fundamentals of Design in Science

Table 9.1 A summary of the appeal, comprehension, retention model, its purposes, and the associations each component evokes

Property	Purpose	Keywords
Appeal	Intrigue and hook the viewer Convey a positive attitude or feeling Define the ethos of the sender	Likable, engaging, beautiful, balanced, harmonious
Comprehension	Help the viewer understand Show the facts	Educational, instructive, easy to understand; good overview; simple to identify individual components and their relationships
Retention	Help the viewer remember	Pedagogic, memorable, noteworthy, enduring

- **Appeal**
  The audience should be motivated to engage, driven by curiosity, interest and fascination.
- **Comprehension**
  The information should be presented in a way that allows the viewer to clearly understand the content.
- **Retention**
  The content should be communicated in a manner that makes it easy to remember.

9.3.2 The CARP Model

The second model is a standard in the design world, made popular by Robin Williams in the classic *Non-Designer's Design Book*.[5] It was also used by designers such as Garr Reynolds in *Presentation Zen*.[6] According to Williams, graphic designers should be guided by four principles, summarized by the acronym CARP[7]:

- Contrast
- Alignment
- Repetition
- Proximity

For an extended version of the model, you can add Space, Balance and Movement. Let's take a look at each principle.

---

[5] Williams (2015).
[6] Reynolds (2011).
[7] You can probably come up with an unprintable alternative.

Contrast
Differences should stand out. Think: tiny vs. huge, blood red vs. snow white, geometric vs. hand-drawn, and so on. From a psychological viewpoint, this corresponds to the principle of salience (see Sect. 5.4.2).

The power of contrast:

- Vivid colors guide the viewer and add meaning—either through their own vibrancy, or by creating unity between elements.
- Strong variation in font size demonstrates visual hierarchy.
- Formal elements balance hand-drawn and quirky elements, creating an impression of energy rather than chaos.
- Over-sized elements, like a large illustration, hook the viewer and stimulate curiosity.

Alignment
The entire layout should be based on an (invisible) background grid, giving everything a structured and purposeful look. The grid also guides the viewer's eye, facilitating visual flow.

The power of alignment:

- Alignment communicates "organization" and indicates that something has reached a final version; unlike a hand-drawn sketch where elements are arbitrarily placed.
- Apart from providing professionalism and a sense of order, alignment guides the audience on how different elements fit together.

Repetition
Repetition works well not only for shapes, fonts, sizes, and colors, but also for the organization of the layout. The simplest way to apply repetition is to create a style sheet for a document or web page: consistent text format, three levels of headers with the same font, size, and color, and so on.

The power of repetition:

- Consistently repeated formatting helps signpost different sections of a presentation by making the design more predictable.
- Repetition also reinforces brand identity (think: *ethos*), and creates a consistent experience for your audience.

Proximity
Items that belong together should be grouped together.

The power of proximity:

- Proximity helps users understand the relationships between different parts of a design. For instance, icons should be placed next to relevant text, or a call-to-action button near related content.

- Effective proximity simplifies navigation, reduces cognitive load, and ensures a cohesive layout.

To these, we can add a few more principles to consider design in greater depth.

Space
The amount of empty area around elements and the distance between them.

Balance
How elements are arranged across the layout, and the distribution between "heavy" and "light" elements.

Movement
The illustration of cause and effect, sequential actions, or steps in a process, often important when explaining mechanisms or processes.

How essential these design principles are becomes apparent when we consider them through a negative definition—their absence. Take a look at Table 9.2.

Table 9.2 Describing the CARP model through negative definitions

If you DON'T follow the principle of…	The resulting impression will be…
Contrast	… dreariness; lack of character and boldness
Alignment	… messiness; carelessness
Repetition	… incoherence; lack of unity
Proximity	… chaos; confusion
Space	… crammed; unappealing
Balance	… imbalance; distortion
Movement	… lack of dynamics; lack of perceived motion

9.3.3 Zen Faulkes' Similarity and Contrast Model

Scientist and communications trainer Dr. Zen Faulkes suggests a minimalistic design model in his book *Better Posters*.[8] According to Faulkes, many design tasks boil down to two directives that stand in contradiction to each other:

- Make stuff appear the same.
- Make stuff appear different.

---

[8] Faulkes (2021).

His model corresponds with the aforementioned CARP model in the way described in Table 9.3.

Table 9.3 The alignment of the CARP and similarity and contrast models

CARP	Similarity and contrast
Contrast	Different!
Alignment	The same!
Repetition	The same!
Proximity	Arranged together = the same! Arranged apart = different!

The principle of "same but different" is always useful in communication. It means that the design follows some kind of cultural convention and professional standard but includes interesting and engaging deviations. At first glance, viewers should know exactly what they are looking at. Immediately after, they should notice that this is something out of the ordinary.

9.4 Guidelines for Reluctant Visual Designers

We'll start with a quick set of robust guidelines that will help you produce slides, posters, and publication figures which do their job—and support, rather than obscure, your content. Later in this chapter (as well as in Part III), we will go into more detail with specific visual design you're most likely to create.

Use a template
A template can save you a lot of time by eliminating the need to make decisions about colors, fonts, positioning, and alignment. The default option is to turn to your own organization and see what they offer. Sure, the color palette may not represent your favorite colors, and the layout may not be perfectly tailored to your needs—but it will likely do its job, allowing you to focus on content, structure, and delivery.[9] If you're ambitious about style and aesthetics, you can, of course, create something of your own.

A problem with templates, though, is that they often contain visual elements that may get in the way of making it 'your own'; meanwhile, creating something yourself may take up too much time. While we appreciate the fine-tuned layout, synchronized colors and elegant typography, we are not that keen on distractions like logos and date stamps. However, with a bold approach, these can—in most cases—be managed.

---

[9] In addition, the use of templates brings stability in stressful situations as you don't have to fear being judged for your design choices—just blame the ugly colors and small fonts on your university's communication department and go on discussing the content!

9.4 Guidelines for Reluctant Visual Designers

Remove graphic junk

The concept of *chartjunk* was introduced in the 1980s by data visualization guru Edward Tufte. The main message is simple: remove all visual details that only serve as distractions for the viewer. We strongly recommend applying this philosophy to graphic design in general.

An example of graphic junk is all the distracting stuff often included in generic slide templates: logos, taglines, copyright notices, dates, slide numbers, and so on. The cleaner you make your slides, the more you focus the viewer's attention on exactly what you want them to look at.[10]

Keep it simple

Go for minimalism. Let yourself be inspired by the classical Neapolitan dish, pizza Margherita, which consists of four things: wheat dough, tomatoes, basil and mozzarella. Minimalistic, yet very tasty! In light of that, use just three colors and two typefaces for your content—or even fewer (see Fig. 9.1). Make the background either solid white or solid black.

Fig. 9.1 A demonstration of the Pizza Powerpointo recipe—keep it simple!

In general, if something in your design feels disharmonious or unbalanced, it is often better to *remove* elements than to *add* them. Aligning and grouping objects carefully is usually more effective than adding frames and lines. Remember the words

---

[10] Joanna must make a note that there are instances where some of this junk MIGHT be useful— for instance, if you are lecturing, audiences find taking notes easier when the slides are numbered; at conferences, including contact details or social media handles on the slides enables a parallel discussion online. Still, make these additions a conscious choice, not a default.

of the French author Antoine de Saint-Exupéry: "Perfection is achieved not when there is nothing more to add, but when there is nothing left to take away."[11]

Make it BIG
Your audience should be able to see and read everything without strain. There is nothing more frustrating than a pixelated figure caption on a poster or an illegible presentation slide (Joanna is the sort of person who always sits in the front row at talks, and even then she sometimes struggles to read the slides—ugh!). Also, using a big font and large images will prevent you from cluttering your visual with too much information.

Use full-size stock photos
A very simple way to create elegant PowerPoint slides or social media graphics is to find a nice stock photo and simple but stylish typography (see Fig. 9.2). This kind of slide was made popular by Garr Reynolds, the designer behind the concept of *Presentation Zen*.[12] Websites like Pixabay, or Sarang's personal favorite, Unsplash, offer royalty-free stock photos. Just make sure to credit the photographer in your work.

Fig. 9.2 A quick, elegant and impactful slide made with a beautiful stock image and minimal text. [*Photo by* Michal Lawrenin on Unsplash]

Use artificial intelligence
Over the past several years, artificial intelligence has made significant strides in the design space. You may already have seen or used it, whether through Microsoft

---

[11] "*Il semble que la perfection soit atteinte non quand il n'y a plus rien à ajouter, mais quand il n'y a plus rien à retrancher.*" From the philosophical autobiography *Terre de Hommes* (1943).
[12] Reynolds (2014).

9.4 Guidelines for Reluctant Visual Designers

PowerPoint suggesting design changes to your slides, or the Content-Aware Fill tool in Photoshop.

AI enhances efficiency, and often opens up many different creative possibilities. Input an art prompt into an AI design generator and explore! But be cautious—avoid using AI to present images of actual biological processes or phenomena, as they might not be accurate (even if the difference can only be spotted by a specialist). In a now notorious example from February 2024, the peer-reviewed scientific journal *Frontiers in Cell Development and Biology* retracted a publication that contained nonsensical images generated by AI, including one of a rat with what looked like gargantuan genitals—giving the term "cock-up" a new level of meaning.

To be on the safe side, use AI primarily for design purposes. Of course, as with any image or reference found online, always note that the visual was created by AI.

Always give credit. ALWAYS.
The expression to "give credit where credit is due" isn't just about ethics or professional integrity—it's also about building trust and showing appreciation for others' great work. Giving credit is also a protective measure: clearly indicating which ideas are yours, and which someone else's, helps you avoid plagiarism (and lawsuits).

Use hand-drawn figures
In his book *The Back of the Napkin*,[13] business communicator Dan Roam demonstrated how simple, hand-drawn figures can be very effective. A bouncy stick figure with some pizzazz wins the hearts and minds of an audience more effectively than sterile clipart. Aim for simplicity, personality, and—of course!—humor.

Imitate
If you see something you like, go ahead and create your own interpretation of it. Just make sure to modify every element so that no one can claim that you're a copycat (that is, infringing copyright laws). There is a fine line between "finding inspiration in something" and "stealing"—do not cross it.

Avoid being an amateur
The mistakes below are some of the most common among non-designers. By avoiding them, your work will look much better.

- **Don't suffer from *horror vacui***
  Beginners in graphic design often have a tendency to fill the entire surface with detail. Not only does this clutter the design, but—more importantly—makes it harder to guide the viewer's attention to your key messages (Fig. 9.3).

---

[13] Roam (2013).

Fig. 9.3 Getting key messages across is easier with a smaller amount of strategically chosen text. Filling a page completely risks the reader glossing over it and missing the most important points. White space is your friend

- **Do not use borders**

  Beginners who are not happy with a layout often believe that borders can help solve the problem. Usually, this only results in a cramped design (Fig. 9.4).

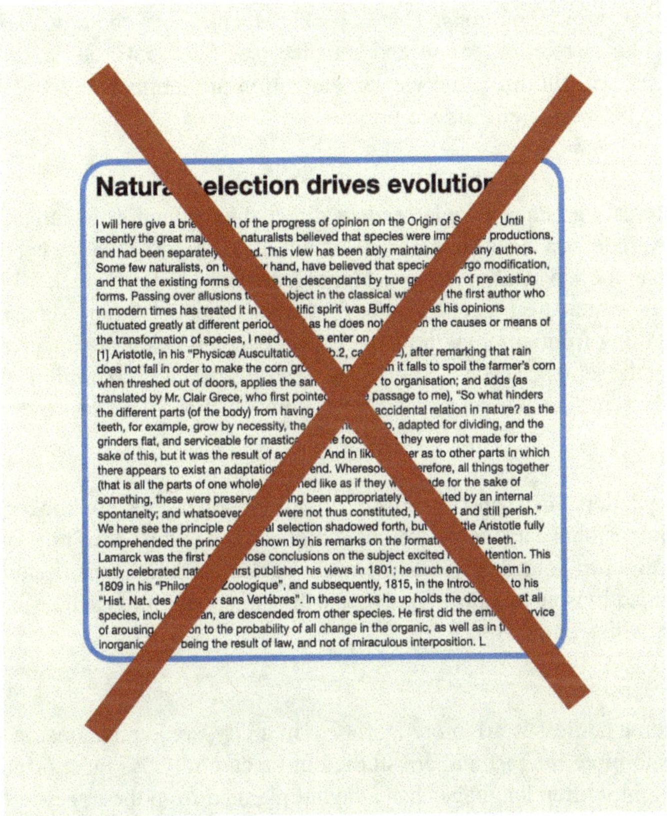

Fig. 9.4 Borders make a cramped design even more cramped

- **Do not use centered text**
  Centered text is only useful for special kinds of print materials, like diplomas, menus, and wedding invitations.

9.5 Visual Hierarchy Helps Us Organize Information

Graphic design is an art that takes a lifetime to master. It's not just about making things look good, but also about making them easy to decode, understand and memorize. Consider a spread in a book with text, illustrations, and photos. Obviously, the content itself carries information and meaning, which should be conveyed as clearly and coherently as possible. But—on a meta level—the layout and the (optional) graphical decorations also convey a message.

A concept most people intuitively understand is visual hierarchy. When we look at a front page of a newspaper or a promotional poster, we can usually identify what the sender finds most important: big, bold headlines shout for attention, bright colors convey strong emotions. This is basic perception psychology, explained by principles like directed attention and salience (see Sect. 5.4). By learning to use a set of cues, we can direct the viewer's attention more effectively and define the relationship between elements we present.

Size

BIG is interesting, since people tend to direct their attention to larger items. In typography, there is a clear hierarchy: the main title is bigger than the header, which is bigger than the sub header, which is bigger than the body text, which is bigger than the captions, which are bigger than the footnotes. Scientists often use acronyms to denote anything from complex proteins to processes, making the uppercase letters stand out in the sentence. Matt Carter in *Designing Science Presentations*[14] suggests lowering the font size of acronyms by 1 or 2 points to make them less overpowering.

Weight

Bold, heavy letters, lines, and icons draw attention; objects that are thicker, textured, or three-dimensional carry more weight. Boldness and texture aren't the only things that add weight to your design—sometimes, packing multiple small elements into a space can add visual weight. An object will also appear more salient if placed in negative or white, empty space.

Position

Just like some addresses are more attractive in a city, some positions on a page or screen create more interest and are studied more carefully. As most printed matter in left-to-right written languages has a layout planned from the upper left, we tend to look in this corner for entry points. On the other hand, in promotional posters or oil-paintings, the center of attention is the center of the image. This is old journalistic knowledge, often reflected in the price of different advertising positions. The expression "Above the fold" indicates that the top of a page is perceived as more important than the bottom.

Grouping

Things that are related together ought to go together, stated German psychologists in the 1920s. They developed the ideas of "gestalt grouping," theorizing how people organize elements into visual groups based on certain principles. These principles are well-utilized in logo branding.

Grouping is often understood as a **proximity** feature—which is true! Even if elements differ in shape, color, or size, sheer nearness can present a figure. For example, the elements of the Apple logo form the shape of an apple, and IBM's horizontal bars create letters. Similar items are meant to be together, like the interlocking C's of Coco Chanel or the interlocking circles of the Olympics. When similar design features are put together, they function as one unit.

---

[14] Carter (2013).

Grouping also accounts for **continuation**, compelling the eye to move from one object to the next in one, continuous swoop. Think of the Nike swoosh or even the classic Coca-Cola logo. Our eyes follow the curvatures so naturally.

> **Text has a funny way of changing the interpretation of visuals**
> As we're discussing the relationship between visuals and words, it is worth mentioning that text can strongly influence how a photo or an illustration is interpreted. A header or caption can easily direct the viewer's attention, as illustrated by the examples in Fig. 9.5.
>
>
>
> **Fig. 9.5** Note how text can change the interpretation of an image. For best effect, look at one photo at a time. [*Photos by* Zack Minor and Thomas Lefebvre on Unsplash]

9.6 Making Typography Work for You

There is nothing magic about typography, even if it may seem so when you talk to aficionados. Many design professionals, especially art directors,[15] can become over-enthusiastic about what's actually possible to communicate through fonts.

Typography is about designing, selecting, and organizing letter forms on a page or screen to achieve your communication goals. The primary purpose is to make the reading process easier by optimizing legibility and clarity. But the reading part is just the beginning. Text on a page or a screen (or a billboard or a sign or a

---

[15] In advertising, a professional designer who focuses on core ideas and visual aspects.

logo …) has personality. It conveys both ethos and pathos. Typography can help the author appear formal or informal, traditional or trendy, classy or unrefined. It can also convey feelings, like "bouncy," "sunny," "depressing," or "stressed." Additionally, it can suggest cultural connotations and associations, such as graffiti style, the wild west, military, high fashion, science fiction, 50s nostalgia, and so on.

Today, most people associate typography with digital fonts found in Microsoft Word, but this nuanced craft has a long history. From Egyptian hieroglyphs and Chinese calligraphy, typography spans thousands of years. Its purpose has not changed much since its inception: arranging letters into words into sentences in a way that is aesthetically appealing, legible, and readable.

Note: Legibility refers to how easily a reader can distinguish and recognize the individual letters and words in a text. It is thus mainly determined by visual design, specifically typography. Factors such as font style, size, alignment, and line spacing help audiences clearly perceive the written words.

On the other hand, readability is about how easily people can read and move along the sentence. Readability is the shared responsibility between the writer and the designer. The writer ensures the content is structured and well-arranged, and the designer enhances the flow through effective typographical design. Both are crucial.

Choosing the right font: serif versus sans-serif

When drafting reports or essays during your secondary education years, you might have been told that your work must be either Times New Roman or Arial, in either 11-point or 12-point font size. As you looked at your options, you may have noticed that between the two font styles, Times New Roman has small lines at the end of the main strokes, while Arial does not. The small lines are "serifs," and the font typeface without serifs is called "sans-serif," from the French word *sans*, meaning "without" (Fig. 9.6).

Serif   Sans-Serif

Abc   Abc

Fig. 9.6 The differences in serif and sans-serif fonts. Note the serifs—or their lack—at the end of the main strokes of the letters

Serif fonts are great for printed text because the serifs quietly guide your eyes to smoothly go from one letter to the next. In fact, the serif typefaces were created for typewriters and printers to mimic handwritten words. Due to their high readability score, serif fonts are used in most journals, books and magazines. Common examples include Garamond, Times New Roman, and Georgia.

Sans-serif fonts exude modernism and sophistication with their simplicity and minimalism. They are preferred for digital formats—think social media, websites, blogs—because serifs can be hard to discern on low-resolution screens. For scientific posters or presentations, sans-serif fonts are preferable as they are easier to read from a distance or in small sizes. For scientists, sans-serif fonts should be the go-to for all things non-manuscript. This book uses Times (a serif font) for the body text, and Myriad (sans-serif font) for titles and headings.

Now, something to keep in mind when selecting typefaces is *appropriateness*. A typeface's appropriateness is based on the desired aesthetics and your preference, but also on the typeface's original design intent and mood. Design intent refers to what the typeface was originally created for. You can discover this with a quick web search. For instance, Microsoft-developed Comic Sans was inspired by comic book lettering. It was meant to introduce young users to computers, and to be used in informal and fun situations, as well as cartoon speech bubbles. Consequently, Comic Sans isn't going to be suitable for the body text of a book, or for a large conference.

Often, the appropriateness of a font depends on the mood it creates, that is, the dynamic mishmash of aesthetics, readability, and the perceived text. Specialty fonts can convey your desired mood, especially in places where you want to take communication to another level. As they can grab a lot of attention, we recommend you use them sparingly, in instances where the take-home point is important. For example, the Indiana Jones franchise uses a style similar to the font Adventure for its logo, signaling excitement and daring escapades. In contrast, the New York Times's logo uses Engravers' Old English BT, which, through its resemblance to a medieval script, exudes the persona of authority and majesty.

When creating a mood, try to think of the exact opposite of the mood you want to create. If you're struggling to identify one, you may want to adjust your typeface choice for a stronger impression (Fig. 9.7).

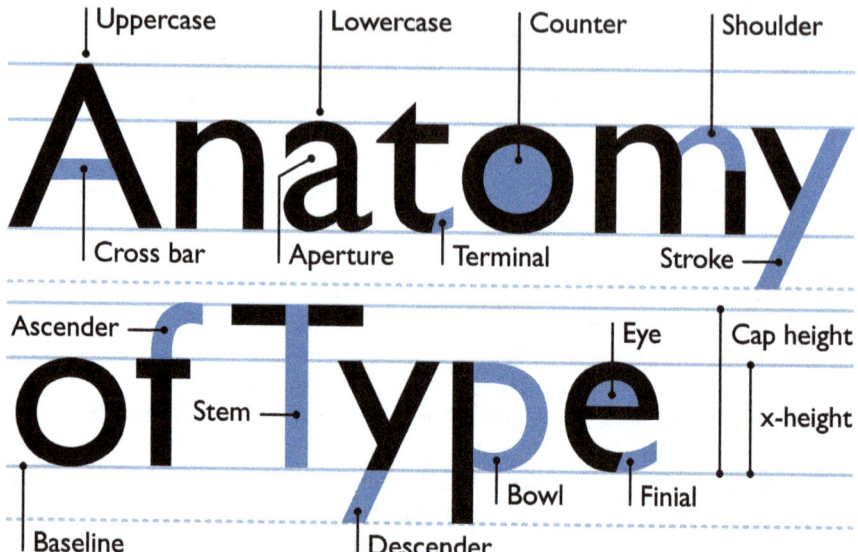

Fig. 9.7 Take a look at the different components of fonts, and how changing small details changes the mood of a visual. [Modified from https://www.interaction-design.org/literature/topics/anatomy-of-type]

Choose combinations of fonts for clarity and readability

Instead of randomly surfing for fonts, get to know a couple well. This is especially useful if you're on a tight deadline. While you might lose some personal style, trusting professional designers' choices won't steer you wrong. Find inspiration in presentations you've found aesthetically pleasing, or via websites like "Fonts In Use"[16] that showcase professional use of fonts across different industries.

When selecting classic neutral serif and sans-serif combinations, make sure that they are complete. A complete typeface is one that is, well, complete. Often, free fonts have incomplete typefaces, with missing serifs on lowercase letters, or lacking full sets of ligatures (which is what happens when two letters combine together to form one glyph). However, you don't need to break the bank and buy a completely new font. Familiar fonts may end up being the best choice.

Further, choose a typeface that has all the numbers, diacritics (like ó, ł, ż), Latin letters *and* non-Latin letters needed for your work. When it comes to numbers, ensure the number 1 obviously looks like a one, not the letter I.

Create your visual style with fonts—but avoid cliches

Sometimes you'll see a font style that immediately appeals to you, for unknown reasons. Trust your gut—but make sure you do it *well*. Your selection should be readable, legible, accessible, and fit the intended mood. Something to avoid however,

---

[16] Fonts In Use (2024).

are cliches connected to the fonts. Are you always selecting Papyrus because your topic is about an ancient civilization? Or are you using Impact for everything that has, well, impact?

Avoiding trite fonts will keep your presentation evergreen, making future edits easier and quicker. Consider using an extended type family font, which includes both serif and sans-serif versions, and is therefore considered complete. This allows you to maintain a consistent mood and aesthetic with just one font.

9.7 Choosing and Managing Colors

In design, very few choices are as important and powerful as the selection of color. Color can draw attention to specific areas or evoke reactions, or stir emotions in your audience. When used correctly, it can enhance comprehension by highlighting specific data. However, if used improperly, color can distract, create the wrong reaction, or worse, make your design unreadable.

To understand how to use color, let's take a brief look into color theory.

Colors on the color wheel can be roughly divided into two categories: warm and cool (see Fig. 9.8). Warm colors include shades of red, yellow and orange, while cool colors cover blue, green and purple. Each color has a distinct emotional personality, and affects the viewers' feelings and perceptions (Fig. 9.9).

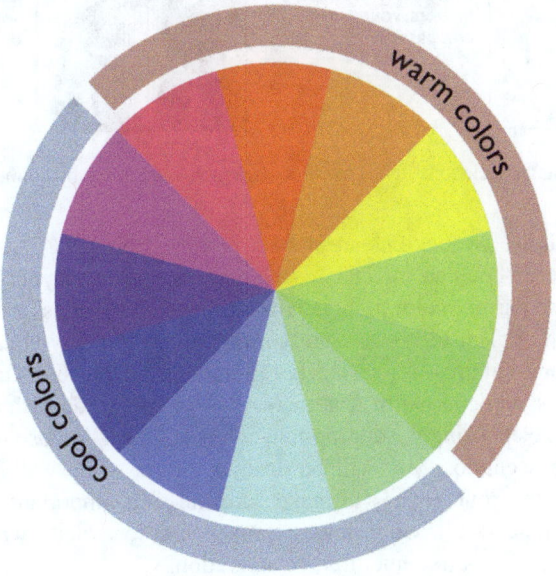

Fig. 9.8 Warm and cool colors on the color wheel

Fig. 9.9 Emotional associations linked to particular colors—the list is by no means exhaustive

These emotional associations are not set in stone, though. A simple change in saturation, tone or hue can affect how the audience takes in your content. Also, bear in mind the current color associations with specific groups or campaigns; for example, pink is linked to breast cancer awareness. This may change depending on the region and culture—for instance, throughout Africa, red is associated with AIDS awareness, while in North America it is often connected with the Red Cross.

We emphasize choosing the appropriate colors so much because in most science communication scenarios, you will be dealing with a white wall, a white page, or a white screen. Your goal is to make sure the data, important facts, or take-home messages stand out. Choosing an awesome color palette will be aided by the knowledge of terms like hue, tint, or saturation.

Hue is the most basic, purest form of a color. When we describe an object as "red" or "blue," we refer to its hue. When it comes to hues, we have **primary**

9.7 Choosing and Managing Colors

colors: red, yellow and blue. They cannot be created by mixing other colors. **Secondary colors**, orange, green and purple, are made by mixing two primary colors (see Fig. 9.10).

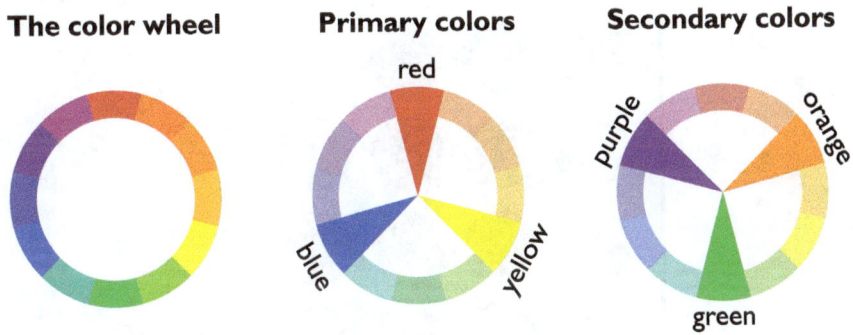

Fig. 9.10 Primary and secondary colors on the color wheel

If you delve into any design guides, they indubitably will include a section expanding on color saturation. They will split hairs between the difference of the terms "chroma" and "saturation," but practically speaking, both terms are used interchangeably. We recommend keeping things simple and just focusing on saturation, as we have done here.

Saturation is the intensity or brilliance of the color. When the hue becomes more and more desaturated, the color tends to look gray (see Fig. 9.11), while a fully saturated color would be a complete hue.

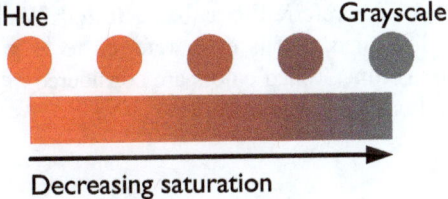

Fig. 9.11 The hue becomes more gray with decreased saturation

A color's **value** is its dimension of lightness or darkness. When the saturation of a color is at its lowest, the value of the color converts to **grayscale**, where white has the lowest value and black has the highest value (see Fig. 9.12).

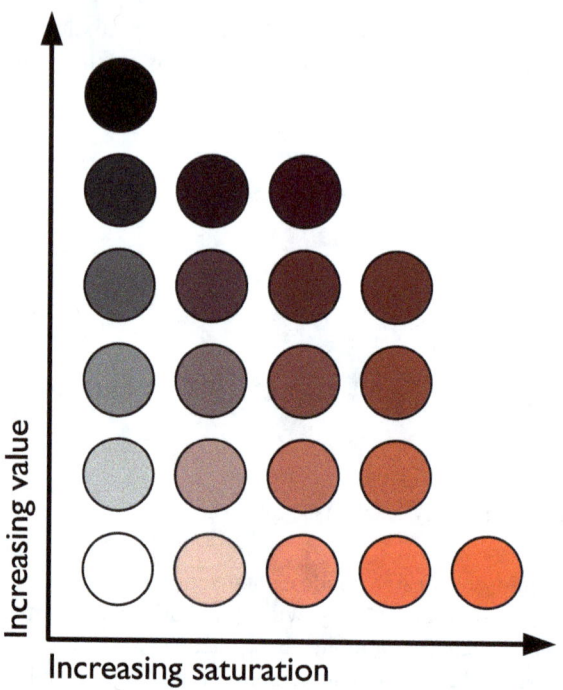

Fig. 9.12 The effects of increasing the saturation and value of a color

Adding black to a color results in a darker, more intense or dramatic effect; the color mixed with black is referred to as a **shade**. When white is added to the hue—no black or gray—it is referred to as being **tinted**. With increased tint, the color is immediately lightened, and is often referred to as pastel. When gray is added, it is referred to as **tone**. Toned colors are considered more pleasing to look at because the gray softens their brightness (Fig. 9.13).

9.7 Choosing and Managing Colors

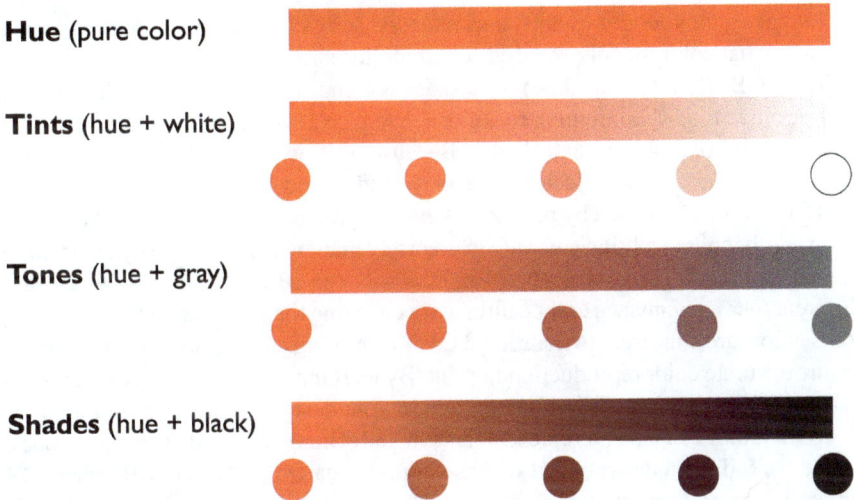

Fig. 9.13 A visualization of the effects of tints, tones and shades on a hue. The sample on the left is 100% base color; the percentage of white, gray and black increases towards the right

These distinctions matter. Lightening or darkening colors adds specific atmosphere or emotion to the personality of the hue. Plus, if you're looking to use more than one or two colors, depending on the shade, tint or tone, you can create contrast that captures attention.

What is the difference between RGB, CMYK, and HEX values?

Before we select any color palettes, we must understand the different models used in design to achieve the most consistent and accurate color representations. These models vary depending on where our design will be used. For instance, papers, flyers, or posters are designed to be printed; presentations will likely be projected onto a wall or shared on a screen, and your personal website can be accessed via laptops, tablets and smartphones. Each of these spaces utilizes color, and each color space has its own unique characteristics (Fig. 9.14).

Fig. 9.14 The color space you pick will depend on the context in which you use it

To make sure your colors appear as intended across different mediums, you need to know what color models are used for which formats.

The **RGB** (**R**ed–**G**reen–**B**lue) color space is commonly used in digital imaging. It is an additive model, combining red, green, and blue at values ranging from 0 to 255 to create a myriad of hues and shades. Because of its excellent ability to reproduce color, it is widely used in electronic displays and digital devices.

RGB is also often used by photographers and graphic designers in Adobe PhotoShop or other photo editing software to precisely manipulate an image by controlling the values of red, green, and blue. Professionals can adjust color imbalances, add different tones, or enhance colors, ultimately elevating their designs. Before printing, RGB colors are converted to traditional CMYK printing colors (mentioned below) to ensure accurate color reproduction in print. By learning just a few things about RGB values, you can achieve higher control across your designs, prints and photography.

RGB values can also be represented as **hexadecimal color values**, or hex codes. These are 6-digit codes prefixed with a # sign, formatted as #RRGGBB, where RR represents red, GG represents green, and BB represents blue. The values, written in hexadecimal (hence the name) range from 0 through 9 plus the letters A through F (representing 10 through 15), determining each color's intensity. For instance, #FF0000 is pure red, with the red component at its maximum value and green and blue at their lowest. Sarang's favorite color in hex code is #6B8D53, which is a light, jade-like forest green, whereas Joanna enjoys the vivid coral coded with #FF5050.

Hex codes are compatible with HTML or CSS coding languages, making them ideal for web design.

The **CMYK** (**C**yan-**M**agenta-**Y**ellow-**BlacK**) color space is most prevalent in printing. It blends cyan, magenta, yellow, and black (also called key) in varying percentages (1–100%) to reproduce desired colors. CMYK is not an additive model, but a subtractive one, meaning that it begins with white and subtracts color to form an image. Unlike the RGB system, the CMYK color model cannot reproduce all shades of color due to ink-light-surface interactions, such as ink absorption, which depends on the color and type of paper.

While CMYK can be viewed on a screen, the colors may differ from the printed material because screens use RGB colors. A workaround is to calibrate your device[17] to ensure the best accuracy. Use these constraints as guideposts when selecting your color palette, especially if you plan to print your designs.

How to choose a color palette

When you open Microsoft PowerPoint or Apple Keynotes, you'll find pre-designed file templates with suggested color palettes. Alternatively, you can use online palette generators like Canva Color Palette Generator, Coolors, or Colormind, which offer versatile and trending color palettes.

---

[17] A color test is often built in-device in the settings to check if calibration is needed (or you can do a check online). Then you can make the adjustments in the control panel and the system will walk you through the process—it is very easy.

9.7 Choosing and Managing Colors

Another great way to create a color scheme is to use colors from a photograph. You can do this online or in an app—just upload an image, whether it's a stock photo or your own, and hey presto! The color tools will spit out a palette for you. These apps will give you the RGB, hex codes, and the CMYK values, so you can make any conversions easily. Apps like Adobe Capture CC even let you pick a mood that will alter the generated color scheme.

But I really want to create my own color palette!
If you prefer to create a custom palette, stick with three to five colors that fit the message, vibe and style you want to convey. Most of your colors should be neutral, like black, white, gray or brown; you'll use these for the majority of your content. Then, pick one or two "accent colors"—like bright red or fuchsia pink—that pop. They'll provide the impact when you want to emphasize parts of your work.

For ease, we recommend you start with traditional color schemes (see Fig. 9.15). A tried and true color scheme is the **monochromatic scheme**, where the same hue is used but changed to a different tone, shade, or tint. Alternatively, one of the easiest schemes to create is the **analogous color scheme**, where three adjacent colors on the color wheel are selected to make a palette. A little more tricky to use are **complementary color schemes**, which is when you choose colors from the opposite sides of the color wheel, and expand it by incorporating different tints and shades. While these look appealing at first, you may find that there is a clash when actually using the colors in practice. So if you're drawn to complementary colors, make sure to add a transition color as well.

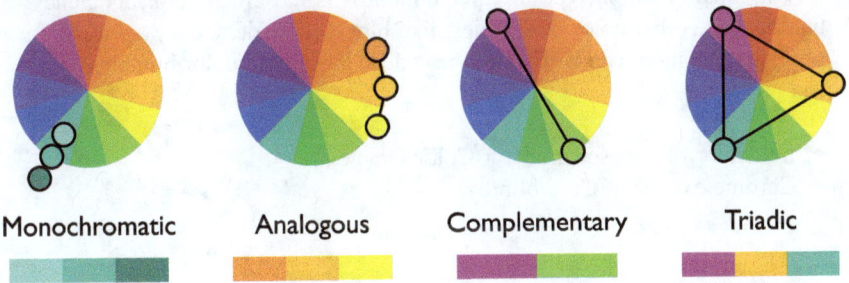

Fig. 9.15 Samples of color palettes that follow the monochromatic, analogous, complementary and triadic schemes

If you're looking for a diverse set of colors or if your work has a lot of moving components, opt for the **triadic color scheme**. Composed of colors equally spaced around the color wheel, one color can act as a neutral, while shades, tones and tints of the other two colors can be used to make information stand out.

Your color palette should also reflect the mood or tone you want to express. Cool grays, black, and white often project modern tones, while warmer browns and rustic reds suit traditional designs.

Choosing accessible colors

Not everyone perceives color the same way. Globally, there are about 2.2 billion people with some sort of visual impairment that affects color perception. This often makes it difficult—or even impossible—for them to understand information signposted by colors alone. It's not just for those who have visual disabilities—as we age, the changes in our vision may hinder clear reading. Plus, anyone can struggle with low-resolution screens or poor lighting conditions. Therefore, accessible colors are hugely important for ensuring usability.

Step 1: Make sure your colors have sufficient contrast

Choose color combinations with enough contrast so that words and images are easily distinguishable. Most color palette generators have color contrast checkers, and some brands have palettes tested for accessibility (for instance, the University of California San Francisco IT department has such a palette).[18] For websites, try to meet the Web Content Accessibility Guidelines (WCAG), which require a minimum contrast ratio of 4.5:1. We recommend using an online color contrast checker like

- Snook, or
- Chrome extension WCAG Color Contrast Checker.

Step 2: Check for color perceptibility

The most common type of color blindness is red-green color deficiency, but there are other color disabilities, like blue-yellow deficiency, as well. To see if your content can be readily viewed by those with color blindness, you can use:

- free color blindness simulator Color Oracle
- Chrome extension Colorblindly.

References

Agero UX Team. (2023, November 29). *Vad är design och varför är det så viktigt (What is design and why is it so important)*. Agero. Retrieved June 3, 2020, from https://www.agero.se/blogg/design

Carter, M. (2013). *Designing science presentations: A visual guide to figures, papers, slides, posters, and more*. Academic Press.

Faulkes, Z. (2021). *Better posters: Plan*. Pelagic Publishing Ltd.

---

[18] UCSF (2024).

References

Fonts In Use. (2024). *Fonts In Use—Type at work in the real world.* Fonts in Use. https://fontsinuse.com/

Kosslyn, S. M. (2007). *Clear and to the point: 8 psychological principles for compelling PowerPoint presentations.* Oxford University Press.

Lankow, J., Ritchie, J., & Crooks, R. (2012). *Infographics: The power of visual storytelling.* Wiley.

Reynolds, G. (2011). *Presentation Zen: Simple ideas on presentation design and delivery.* New Riders.

Reynolds, G. (2014). *Presentation Zen.* Presentation Zen. Retrieved June 15, 2020, from https://www.presentationzen.com/

Roam, D. (2013). *The back of the napkin (expanded edition): Solving problems and selling ideas with pictures.* Portfolio.

UCSF. (2024). *Color.* UCSF Brand Identity; The Regents of the University of California. Retrieved June 17, 2020, from https://identity.ucsf.edu/brand-guide/color

Williams, R. (2015). *The non-designer's design book: Design and typographic principles for the visual novice.* Pearson Education.

Part III
Addressing Different Tasks, Target Groups and Situations

Part III of this handbook addresses a wide range of communication tasks with varying purposes, contexts, and channels. Here, we're using the whole toolbox: speaking and presenting, writing, designing visuals, reaching out to professional partners and the general public, teaching, engaging with social media and networking.

To truly master all these communication tasks will, of course, take years of skill development, practice, and experience. But we hope that these "recipes" will be helpful for you in the here and now when you look for a place to start.

While studying the instructions and the advice, please remember that communication is always most powerful when it is grounded in the key principles we have discussed, for example, a passion for the subject, insight into the audience's needs, and curiosity about the communication methods used by pros.

Finally, we would like to add a reminder that there are no stronger ethos-builders than honesty and authenticity. Learn, study, imitate, practice—and then do it your own way!

Using the recipe book

The remainder of this book is a collection of practical instructions and advice for a number of communication activities.

Starting box

Every "recipe" starts with a box outlining the purpose of the communication activity and some typical situations where it applies. This box also highlights frequent mistakes to avoid.

Intro

A brief introduction summarizes the communication activity and reflects on what you should think about to make the best of your time and efforts. In some instances, the introduction contains references to published work or experience-based know-how.

Step-by-step instructions

These will take you through the planning, preparations and execution from a practical perspective. To get an overview, we suggest you start by skim-reading the bolded opening sentences of each step.

Advice from the pros

Here and there you will find boxes with advice from pros who specialize in a particular area of communication. Pay attention to what they have to say, as it represents years of experience and structured thinking, and send them a thought of gratitude for being generous with their professional skills.

10 Instructions for Speaking in Different Settings

10.1 Introduction

In Chap. 7, we covered the basics of oral communication. With the advice given there, alongside practice and experience, you will build a solid foundation as a speaker and lecturer.

In this chapter, we're taking the next step: addressing more ambitious communication tasks in a broader range of situations. Or, to put it differently, we are adding some further examples of communication activities where your voice, personality, and preparation are your most important assets. Some fast thinking may come in handy—either to dodge unforeseen stumbling blocks or to catch opportunities as they knock on your door.

10.2 The One-Minute Elevator Pitch

> **Purpose**: to quickly, clearly and convincingly convey to other professionals (1) who you are, (2) what you do, and (3) what you offer.
> **Typical situations**: spontaneous interactions during meetings and conferences; ice-breaking during mingling; encounters with potential customers or collaborators.
> **Mistakes to avoid**: getting lost among the details instead of communicating your value; emphasizing the past, rather than the present; forgetting to follow up by engaging in conversation; selling too aggressively.

Many have tried to define what an elevator pitch is, but the concept still appears a bit elusive. Most definitions agree on these key traits:

- It is brief: between 30 s and 3 min.
- It is about you: your project, your product, your company, your business idea.
- It has a purpose: to inform, persuade, sell, impress.
- It seeks a response: the preferred outcome is invitation to dialogue, exchange of business cards, or a LinkedIn connection.

Our flavor of elevator pitch—which we present below—can be summarized like this:

- It is short, precise and interesting. We recommend aiming for 60 s—it corresponds to a little over a hundred words in English, plus a bit of silence to slow it down (and for rhetorical effect). It doesn't really matter whether the pitch lasts 50 or 70 s; we've just found from personal experience that 30 s is too short and 2 min far too long (a very subjective assessment indeed, but bear with us!).
- It is about you and what you eagerly want to communicate to the world: your research, your company, your project, a service, a product or a cause. It is essential that you convey not only what you offer, but why it matters. Meanwhile, the million dollar question for the listener is "Why should I care?"
- The purpose of the pitch is to position yourself as a professional, ignite interest in what you're promoting, and open doors for cooperation, funding and implementation. Above all: your goal is to leave a mark and make yourself memorable.
- The preferred outcome is a continued conversation and a meaningful connection—which may include the exchange of contact details, lunch plans, or a follow-up meeting.

The following protocol is based on the advice of American presentation coach Carmine Gallo, writer of *The Presentation Secrets of Steve Jobs*.[1] This approach has been tested in numerous training workshops and gives a solid, useful foundation. Keep in mind that you might be preparing either a general elevator pitch suitable for most occasions, or a specific one tailored for a particular event.

1. **Go through four basic questions**

    Respond to the following questions. Be spontaneous and don't think too much about the final result.

    I. What do you do as a scientist (researcher, entrepreneur, technician…)?
    II. What problem do you solve?
    III. How is your research unique?
    IV. Why should I (your audience) care?

    Remember, these four questions represent a framework. Feel free to modify them so that they fit your own story (see box below).

---

[1] Gallo (2010).

> **Some variations of the four questions**
>
> I. What do you do as a professional? What does your company/organization do? How would you describe your focal subject in two sentences?
> II. In what way do the things you do help others? What services do you offer? What services does your company/organization offer? How will your solution make the world a better place?
> III. How is your research/business/organization/product/solution different?
> IV. *Nope—no variations for this one!*

2. **Edit the text and keep an eye on word count**

   Now, focus on giving the pitch some flow. Write down your responses and trim to 90–100 words. As you will soon discover, this step may be a tough one, especially if you are the kind of person who tends to include many details or feels uncomfortable with simplifying your explanations. Persevere—you'll come out the other end with something really effective.

3. **Polish the language**

   Once you have a clear, concise, and engaging draft, it is time to read it out loud.

   Does it sound like you? The pitch should mirror your natural conversational style. Avoid fancy words and complex phrasing that might make you stumble.

   How promotional should you be? The long answer to this lies in an ethos/pathos/logos analysis (see Sect. 4.4.1). The short answer is: it depends on the purpose, context, and how comfortable you are with self-promotion. If your purpose is to sell, the context is commercial, and you're happy to promote yourself—brag away! Otherwise, find a tone which aligns with the situation and your personality.

   How creative and fun should you be? Same thing here: it depends on the purpose, the context, and how bold you want to be in challenging preconceived notions. A science academy in Eastern Europe is probably most impressed by something formal and conventional, whereas the audience at a tech unconference in Scandinavia expects something edgy and fun; it's your call. Most importantly, present something that is in sync with your personality.

   Time yourself reading your draft. At your natural pace, it will probably take 40 s or more.

4. **Taste the words**

   Now try to perform your pitch. Does it sound clear, interesting, and engaging? If not, keep editing. Taste the words and the sentences: are there any "rough edges" or expressions that don't sound like you? If there are, keep editing.

5. **Practice**

Now is the time to memorize the pitch. But please remember that it is more important to convey your message with credibility and pizzazz, than to memorize every single word verbatim from your manuscript.

After practicing alone, try the whole thing on your friends and hear what they think.

6. **Internalize, adapt and put to use**

Your pitch is like your phone: something you carry in your pocket and pull out as needed. Always adapt the content and the style to the context; think on your feet and make a quick decision on what's appropriate, convenient and meaningful here and now. Have several versions ready, and whip them out depending on your audience (just as you would tailor your CV to specific job applications).

Some variations of the elevator pitch

Written version

I am a scientific illustrator specializing in publication figures, graphical abstracts, and video animations. I also offer training in the form of live workshops and online seminars. With a Ph.D. in neurochemistry and years of experience in the scientific publishing sector, I not only have a unique understanding of the scientific method—I am also able to deliver a quick turnaround from draft to publication. Most importantly, I feel comfortable in the sometimes peculiar culture of academia. For research groups who choose my services, I act as an in-house visual communicator, offering clear, stylish and cost-effective designs for a variety of channels.

Spoken version

I work as a scientific illustrator. I would say that my specialties are publication figures, graphical abstracts and video animations. Sometimes I also give workshops and seminars—both live and online. If you should ask me what sets me apart, I think it is that—ONE!—I have a Ph.D. in neurochemistry and—TWO!—I have worked in the scientific publishing industry for many years. This gives me a unique understanding of the scientific method, and I also work faster than most illustrators. For my customers, it is very useful that I feel totally at ease in the culture of academia which can be—what-should-I-say?—a bit peculiar sometimes [BIG GRIN]. When I work together with a research group, I become what you could call an "in-house visual communicator," helping them with clear, stylish and cost-effective designs for different channels.

10.2 The One-Minute Elevator Pitch

Short and relaxed version

I am a scientific illustrator and I like to work with publication figures, graphical abstracts, and video animations. I feel at ease with the peculiar culture of academia as I have a Ph.D. in neurochemistry and have worked in the scientific publishing industry for years. My way of working is to become a kind of in-house visual communicator for research groups—they email me a draft, we have a discussion, and then I deliver something promptly.

Conversational version

So what do you do?

I am a scientific illustrator, you know—publication figures, graphical abstracts, video animations and stuff. In addition, I run courses every now and then.

You work with scientists then?

Exactly! It helps that I have a Ph.D. in neurochemistry and have worked in the scientific publishing industry for many years. I really understand what I am doing, and I also work faster than most illustrators.

But why can't scientists just produce their own illustrations?

A lot of scientists do and many of them are great illustrators, especially the younger generation. But frankly, most lab people are better at research than they are at communication. So, in some of the groups I work with, I have become a kind of in-house visual communicator. You know—a one-stop-shop for all their visual design needs, adapted for different channels. It saves them a lot of time!

7. **Bonus use**

As soon as you understand that your elevator pitch is just a starting point for succinct communication, you can edit it in whatever way you want and use it as the following:
- LinkedIn summary
- University web bio
- ResearchGate bio
- Conference bio
- Fact box for interview on a website or in a podcast.

> **Self-marketing may feel like bragging**
>
> Many people, especially the overthinkers and high-achievers, find it uncomfortable to speak about themselves in positive terms. To ease this discomfort, they may find it helpful to lean on feedback from trusted friends, who can help them shine in the presence of an audience.
>
> In our experience, attitudes toward self-promotion differ significantly between cultures. Scandinavia, for example, is cursed by what's called "Law of Jante"—a set of social codes that discourages expressions of individuality and personal success.[2] This is very far from, for instance, the typical North American salesman style that has become a cultural stereotype.

10.3 Marketing an Idea, Project or Product

> **Purpose**: to use a more aggressive method to sell a product, solution or idea, based on sales techniques from trade and industry.
>
> **Typical situations**: negotiations, crowdfunding, promotion of events and campaigns, ticket sales, charity activities.
>
> **Mistakes to avoid**: leaving either party disappointed; taking things too far and damaging the reputation (yours or your organization's); finishing without asking for closure (agreement) or a clear intent ("next action").

If you asked a group of Ph.D. students how they feel about becoming salespeople, the majority would likely find the thought experiment in itself a bit off-putting. In our experience, the mindset and personality of most science people seem to lie far away from that of typical sales representatives.

As a matter of fact, mastering selling techniques can come in handy. Here's why: life is very much about selling stuff. Not just physical products like cars, clothes, or concert tickets, but immaterial things like ideas, plans, and suggestions. Negotiations, for instance, can be seen as a dual sales process: "you want something from me, I want something from you—let's see if we can match these needs while keeping both parties happy."

In Sect. 10.2, you learned how to prepare a compelling elevator pitch. Now, we will turn the temperature up a bit and introduce some effective sales techniques. These can be used not only to sell your old hockey gear at a garage sale, or your

---

[2] In Australia and New Zealand, a similar phenomenon, known as "tall poppy syndrome," refers to successful people being intensely criticized when their peers believe them to be flaunting their success.

10.3 Marketing an Idea, Project or Product

lab's unused PCR machine—they may also serve you when you are presenting a solution to your research colleagues, or pitching a concept in a board meeting.

1. **Define exactly *what* you are selling**

    You can, for example, start with the user story model (see Sect. 3.2.2).
    - **Problem**: what customer needs does your product/solution/concept/suggestion/idea correspond to; what problem does it solve?
    - **Solution**: exactly how does it prove itself useful?
    - **Outcome**: how will things work when your product/solution has been implemented?

    Then, you can go on with the Five Ws:
    - **What** are you selling?
    - **Who** is the customer?
    - **Why** do they need or want your product?
    - **Where** will it be used?
    - **When** will it be used?

    Finally, use the Aristotelian triad as demonstrated in Table 10.1.

    Table 10.1 Aligning your search strategy with the Aristotelian triad

Logos	Ethos	Pathos
• Does the product/solution solve a problem or match the requirements? • Is there an acceptable value for money or return of effort? • Will everyday life be made more convenient by implementing the product/solution? • Do you offer a competitive advantage, something new and unique? • Are you mitigating a business risk, or complying to regulations? • ...	• Do you represent authority and trust? • Do you understand and care about the buyer's needs/problems? • Is the stuff you offer any good in the long run? • Do you have the background and experience to design a sustainable solution? • Do you have a good reputation/track record? • ...	• Is it pretty, cool or alluring? • Will it increase the social prestige of the buyer? • Will it make the buyer feel safer? • ...

2. **Take a look at yourself**

    Are you the right person to sell this product? Do your persona and behavior sync with the values and social expectations of the customer? If not, is there something you can adjust to build trust? Or perhaps you can team up with someone who has a stronger ethos? (Please note that even the fact that you work in a team can enhance your perceived trustworthiness!).

3. **Create a sales funnel**

This is a fundamental concept in sales and marketing. The metaphor implies that potential customers enter a funnel that eventually leads them to a closed deal. A general rule of thumb is that ten initial contacts may lead to three meaningful dialogs, which then may result in one closing. By knowing what to focus on and learning how to interpret feedback, you improve these odds and spend your time more effectively.

The sales funnel
In its most basic form, the process follows the acronym AIDA.[3]

- **Attention/Awareness**: ensure your product/solution is noticed and evokes curiosity.
- **Interest**: pull the customer in by presenting alluring information.
- **Desire**: turn up the heat by throwing in some killer features, and by creating a vision of a better future (the "Outcome" part of the user story). Make it hard for the customer to pull out.
- **Action**: close the deal by laying out a safe and easy way for the customer to say YES and haul up their wallet. To create a sense of urgency, you can set up a "it's now or never!" situation (Fig. 10.1).

Fig. 10.1 The sales funnel consists of four main points: attention, interest, desire and action

---

[3] According to Edward K. Strong, the slogan *"attract attention, maintain interest, create desire"* is attributed to Elias St. Elmo Lewis, who coined it in 1898, later adding *"get action"* to the mix.

10.3 Marketing an Idea, Project or Product

Below are a couple of examples of AIDA in action.

Example 1:
You enter a store where a salesman in a bright orange polo shirt is playing with a cool gadget, looking very entertained. You become curious and can't stop yourself from watching him. He spots you and strikes a conversation. In a casual way he asks you a set of questions about yourself, locking you tighter and tighter into the conversation. As he now knows that your birthday is coming up next week, he offers you a great deal on the gadget: "This is actually our first unit—a demo, really; I can give you 25 percent off!".

Example 2:
PCR is like VCR: an obsolete technology **[Awareness phase]**

With PCXR, MolBio-X introduces a groundbreaking way of simplifying your lab work:

- *Better yields*
- *Lower costs*
- *Simpler handling of primers* **[Interest phase]**

Beta-testing in five top colleges in the USA showed 18% lower cost and 24% shorter processing time than current PCR technology. (Trade News Mol. Bio., October 2024) **[Desire phase]**

Contact us at sales@molbio-x.com for an introductory deal (valid until Dec 31, 2025). **[Action phase]**

4. **Optional: merge your sales funnel planning with other communication activities**

    You can, for example, place the elevator pitch (see Sect. 10.2) and the sales funnel together (Table 10.2).

Table 10.2 Aligning the sales funnel (left) to the elevator pitch structure (right)

Attention	Gimmick, hook, something out of the ordinary... See principle of salience in Sect. 5.4.2
Interest I	*What do you do as a scientist?*
Interest II	*What problem(s) do you solve?*
Interest III	*How is your research different?*
Desire I	*Why should I care?*
Desire II	Make it even more intriguing!
Action	Lay out the path to a decision—make it SAFE, EASY, and create a sense of URGENCY

10.4 Participating in a Panel

> **Purpose**: to offer an audience an interesting mix of facts, explanations, opinion, ideas and suggestions—either as a brainstorm, a scrutiny, or as a dialectic discussion.
> **Typical situations**: conferences—live and online.
> **Mistakes to avoid**: forgetting to make it interesting and engaging for the audience; lack of dynamics and interaction; competing with other panelists for prestige instead of boosting each other.

A panel discussion brings together a group of interesting, knowledgeable people to have a structured conversation about a certain topic in front of an audience. If you play it safe, you rely on established thought leaders and trusted experts in the field. If, on the other hand, you want to turn up the heat, you could add a single maverick or a pair of disagreeing nonconformists to the mix.

- A *good* panel discussion is like a potluck supper, where each participant brings some tasty dish to the table.
- An *excellent* panel discussion is more like a collective, democratic cooking experience where completely new dishes are offered to the audience.

It is quite rare that organizers define a special goal or outcome for a panel discussion. Instead, they gather a group of people who have something substantial to say about the subject in question and hope for it to become interesting naturally. But, as we keep saying, communication is most effective with planning. So, in the spirit of this book, we strongly recommend that you define a purpose for the discussion. Below, we suggest some ideas.

10.4 Participating in a Panel

- A **brainstorm**: when experts and thought leaders gather to generate ideas, visions, and potential solutions.
- A **scrutiny**: when experts from different fields analyze a problem from their unique perspectives.
- A **dialectic discussion**: when disagreeing—at least initially—parties engage in a conversation that ideally leads to a new, shared way perspective—or at least to an understanding that their insights are contributing to a broader, evolving truth.

But enough about organizing a panel: here's a guide to help you prepare if you're invited to participate.

1. **Gather all relevant information**

    This is typically something you should do as soon as you decide to accept the invitation—or rather, while you are still preparing to accept the invitation. If you know your What, When, Where etc., there will be no hidden surprises when the D-Day approaches. During early-stage discussions with the organizers, make sure they write down what they expect from you and send it in an email.
    Throughout your preparations, address a number of questions:
    What, exactly, is the subject and the angle? What is the format? Who are the other panelists? Who will be the chair/moderator, and how will they run things? Should I give a presentation at the start, or am I only supposed to take part in the discussion and answer questions? Will I introduce myself, or will the moderator do it? Will everyone in the room be able to join the discussion: the moderator, the panelists and the audience, or only the panel participants?

2. **Reflect on who is the beneficiary of the discussion**

    The simple answer is: the audience in the room. However, in some instances, the media people present at the event may be the top priority. There could also be other kinds of professionals on whom, for one reason or another, you might want to make an impression: funders, industry representatives, politicians, influencers, and so on.

3. **Get to know the team**

    In the panel, you need to work as a team. You don't need to love each other, and you don't have to agree, but, for the sake of the audience, you should act in a coordinated way and follow some rules of conduct—just as a band or a sports team would.
    Therefore, do all the research you can on the other team members—check their LinkedIn profiles and do a bit of searching, then send them an email and say hi (though without starting anything that could be considered scheming).

If you want to hear their voice and perhaps see their face in advance, ask the moderator if they can organize a video meeting prior to the event.

4. **Prepare to give some good answers and contribute to the audience's experience.**

   Here, you can follow the instructions for preparing for a Q&A (see Sect. 7.5).

5. **Support the chairperson**

   In an old patriotic Swedish song addressed to the King of Sweden, the subjects promise to "make the crown light on his head," which means something like "you are our leader, lead us well and we will try to support you in return." This is a great attitude for a panel member. Give the moderator a phone or video call, present your background, the main messages you want to convey, and your expectations. If there is some practical information that can be useful, now would be a good time to mention it.

6. **Be aware, be present**

   On the day, it could be useful to have a little notepad with your key messages (or perhaps important statistics, or difficult names) written down. However, don't just focus on your own performance—be present in the moment, listen to what the others are saying, and refer to it in your statements. If a fellow panelist makes a valid point, amplify it ("as Dr. Nowakowska mentioned…"). Bring a pen, so you can jot down any thoughts that emerge while the others are speaking, or any multi-tiered audience questions. Give the other panelists time to speak—don't hog the microphone, and keep your answers brief; remember that panels are not lectures.

   Also, if you wish, try to engage the audience. Ask them to raise their hand if they ever found themselves in a particular situation, or perhaps even to give a little shoutout (make sure this interaction is suited to the cultural context you are in).

7. **Enjoy yourself!**

   You have been invited to this panel because you are an expert, and your opinion matters. People have come to hear what you have to say. This in itself should be a pleasant ego-booster—so go out there and enjoy your time on stage.

10.4 Participating in a Panel

Moderating a panel

Three tips from Natalia Osica, founder of 'pro science', the first Polish PR agency directed at academics.

Tip 1: Perfect your panel discussion with an optimal schedule

To ensure your panel discussion runs smoothly, it's crucial to plan how you'll allocate time to each segment. This plan will be your guide to a successful event. Let's assume you have 30 min for your panel. Here's a step-by-step breakdown:

- **Kick-off and introductions (3 min)**: Start by welcoming your audience and introducing the panelists.
- **Wrap-up and final thoughts (3 min)**: Conclude with each panelist sharing a key takeaway.
- **Guest participation**: With three guests, allocate 2 min per answer. This gives you the opportunity to ask three questions to each panelist.

Don't forget to leave some room for your questions and to clarify certain points. Encourage interaction among the panelists for a lively discussion. For longer panels, invite the audience to join the conversation.

Tip 2: Connect with your panelists before the event

Reach out to each participant ahead of time to learn how they prefer to be introduced, what topics they're passionate about, and any subjects they wish to avoid. This is also your chance to share your vision for the panel and confirm logistics.

Send a group email to introduce the panelists to each other. If possible, arrange a brief video meeting. This helps everyone get acquainted, ensuring a smoother and more engaging discussion. Knowing your panelists' preferences will help you direct the right questions to the right people.

Tip 3: Craft a memorable closing question

Prepare a thoughtful question related to the main topic of the discussion. At the end of the panel, ask each panelist to respond in one sentence. This will serve as a powerful conclusion, leaving the audience with a clear and impactful takeaway.

10.5 Giving Longer Talks and Workshops (1–6 h)

> **Purpose**: to inform, educate, or inspire an audience, based on a more extensive curriculum or study material.
> **Typical situations**: university teaching, professional training.
> **Mistakes to avoid**: forgetting to make the audience feel comfortable, interested and motivated; forgetting about physiological needs; poor time management.

When you are asked to give a longer talk—anything from a one hour lecture to a six hour workshop—you need to consider, and be in control of, five things:

(A) **The social system taking shape in the room**

During short talks, your audience will not have time to reflect on the people around them. Instead, they will focus on your message, make a few observations about what's going on in the room, ask a question or two—then leave.

During longer sessions, however, the situation is different. Now the audience members have time to observe each other's behavior, attitudes, and knowledge. They will start forming opinions about everyone, and they will be aware of their own social role and position in the hierarchy. You, the speaker, will not be exempt from scrutiny. Remember: you're not simply a human multimedia automaton, but a part of the social system—with a backstory, emotions and values.

In most cases, people are kind, altruistic, and disciplined enough to let this social interplay be generous, fair and polite. But on occasion the dynamics—if not handled well—may turn into a power struggle in the room, or can make some participants feel alienated. Then your job as a group leader becomes to direct the social system to make everyone comfortable, interested and motivated. This task may be challenging, but it is also very interesting!

(B) **Your audience's attention span and physiology**

Due to the activity of the brain, the pancreas, the kidneys, and other organs, the members of your audience will need to have a sip of tea or coffee and perhaps get a snack or go to the bathroom.

(C) **Your own attention span and physiology**

As you too are one of the humans in the room, you will also need to manage your own blood sugar and your bladder activity.

10.5 Giving Longer Talks and Workshops (1–6 h)

(D) **The pace and flow of delivery**

To keep your audience interested, you must be the director of your own show: there should be a dynamic variation of pace, communication channels, angles, and—if possible—frequent speaker-audience interaction.

(E) **The time**

Poor time management during a ten-minute talk may result in annoyance. In contrast, poor time management during a four-hour session may result in a total collapse of the schedule and skipping over parts of the material you were planning to cover. If people have traveled far to listen to you, this may cause huge disappointment.

With the above points in mind, you can start preparing for your big day.

1. **Begin with the normal preparation steps for a presentation**

   Analyze the task and the audience, define the outcomes and compile your material—see Chap. 6.

2. **Divide the session into parts that match the time schedule**

   The most convenient approach is to let the coffee breaks and the lunch break be the dividers (Fig. 10.2).

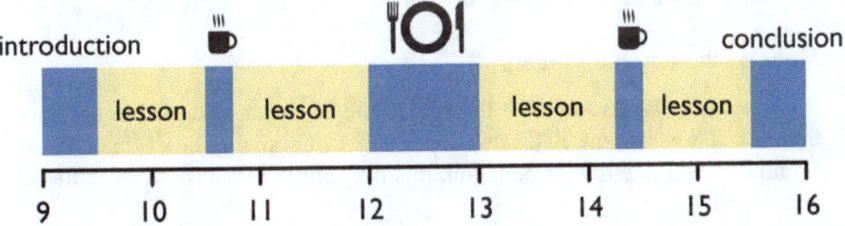

Fig. 10.2 Setting up a timeline before a full-day lecture/seminar/workshop may help you get a realistic grasp of the time you have available. In this interactive workshop, with plenty of time for the intro and the conclusion, the 7-h day contains 4 h and 30 min of effective lecturing and exercise time

- For a day of lectures, where the participants know you and each other, and feel safe within the context, 10 min each for introduction and summing up is enough.
- For an interactive workshop day, where students don't know you or each other, and are facing exercises that put them outside their comfort zone, intro and conclusion may require up to 30 min each.

As you quickly notice, a seven-hour day melts like an ice cube in the Sahara once you do your planning in a realistic way. The normal outcome is four sessions, each between 60 and 90 min in length.

3. **Make a plan for introducing yourself**

   The model is obvious: ethos, pathos, logos (see Sect. 4.4.1). Make them trust you (*ethos*), make them comfortable and confident about working with you (*pathos*), and, finally, give them the information they need about you, your background, and your projects (*logos*).

4. **Make a plan for the social introduction**

   Perhaps a quick, ten-minute chitter-chatter between teacher and group, including a brief interactive online survey (such as Mentimeter, PollEverywhere or Kazoo)? Or why not prepare a thirty-minute icebreaker exercise where everybody has a couple of minutes to present themselves, their pet projects, and their personal values?

   The length and the content of the intro can differ widely, depending on the task, the context, the participants and the subject. For an interactive session or discussion where people need to be candid and a bit vulnerable, it can be worthwhile to spend up to forty minutes, perhaps even an hour, on such an exercise. However, consider whether your exercise scales: while it's perfectly doable to organize an icebreaker that introduces each one of, say, fifteen participants, doing so for an audience of ninety is unfeasible and impractical. Here, an online interactive poll would work much better.

5. **Make a plan for summing up**

   What needs to be conveyed at the end of the day? A call to action or instructions for the next step? A discussion or Q&A? Or simply an evaluation and a summary? Just like the introduction, this may either only take ten minutes or a whole hour.

6. **Stitch it all together**

   While it is a good idea to work on every part of the session like it is a separate presentation, think about what links them, what transitions you are going to make, and how you will signpost your participants towards the next step. What is the flow of the day going to be like?

   Tip: if you are using slides, use different colors to keep the different segments apart—read more in Sect. 12.2.

7. **On the day of the presentation, focus just as much on A–E below as on the content of your talk**

   Let's go through them once again:
   A. The social system formed in the room.
   B. Your audience's attention span and physiology.
   C. Your own attention span and physiology.
   D. The pace and flow of delivery.
   E. The time.

And then you are good to go!

Some tips for your big day

- If you are the kind of speaker who tends to lose track of time, ask for help. Choose a motivated attendee as a timekeeper or encourage the audience just to shout out when a break is due. If neither is likely to work, it might be a good idea to set a timer on your phone.
- Avoid detailing the entire day's agenda. Sometimes you will be forced to skip parts of your material—for whatever reason. Managing expectations is much easier if no one in the audience has been informed about your initial plan.
- Stay away from sweet snacks, especially during the later coffee breaks, as the insulin release they trigger will give you a blood sugar dip. Bring your own slow-carb snacks (e.g., from the hotel breakfast buffet) if the hosts only offer cookies and pastries. Read more in Sect. 7.6.1.
- If your lecture is being recorded, be aware of where the microphone is (ideally arrange a lapel mic, so that you are not tied to one spot). If there are questions from the audience, repeat them, so the people watching online have an idea of what was asked.

10.6 Planning and Executing a Teaching Session

Purpose: to help students understand, learn and remember their subject matter.
Typical situations: planning a class, giving a guest teaching session, improving the structure of presentations.
Mistakes to avoid: failing to keep the students motivated and engaged; failing to reach the pedagogical goals or learning outcomes.

Disclaimer: If you are about to give regular classes within the framework of a teaching institution, the instruction below will not get you far—instead, turn to your university which most likely offers a fitting pedagogical training program. However, if you're giving the odd class, or find yourself accepting to substitute for a colleague, this framework will help you get started.

There are many models and templates for lesson planning. The one we have chosen here originates from Canada and is called the BOPPPS teaching model. The beauty of BOPPPS is that it is a closed-loop feedback model with a focus on teaching interaction and self-reflection. Basically, it is an interactive lesson, bookended with the overview and expectations part at the start, and an evaluation part at the conclusion.

Below is our flavor of the BOPPPS lesson plan.

The BOPPPS lesson structure

B, as in "Bridging in"

The session starts with an activity that sets the tone and introduces the content in an inspirational or intriguing way. If you manage to do this well, the group will be motivated to follow you into what lies ahead.

O, as in "Outcomes"

The next step is to clearly define the outcomes and objectives of the lesson. Since BOPPPS is a closed-loop format, this part should be mirrored by the Post-assessment and the Summary.

The first P, as in "Pre-assessment"

The pre-assessment step serves two purposes. Firstly, it gives you, the teacher, an understanding of your participants: how familiar are they with the subject, how detailed is their knowledge, what is their attitude towards the topic? Secondly, it helps to motivate the students by clarifying what it is that they are about to learn.

The second P, as in "Participatory Learning Activity"

This part is the actual content of the training session. Focus it on interaction—discussion, exercises—and take care to cover the learning outcomes and objectives stated at the beginning.

The third P, as in "Post-assessment"

Did the group learn anything? Now is the time to find out! The simplest way is to run a quiz or let the students write a self-reflection. If you are more ambitious (and if there is enough time), the students may even use their newly acquired skills to solve a problem or do an assignment.

S, as in "Summary"

To conclude the learning experience, summarize the scope and content of the lesson.

10.6 Planning and Executing a Teaching Session

The BOPPPS structure makes it easy to quickly roll out something that works adequately, but please bear in mind the famous slogan of the board game Othello: it takes "a minute to learn but a lifetime to master." With that at the back of your mind, have a read through our preparation steps below.

1. **Define the task**

   You know the drill: who, when, where … and so on (see Sect. 6.2). The most essential question for your BOPPPS planning will be: *"What is the goal of this communication activity?"* This will define the "Outcomes" part.

   Another important question is, of course, how much your students already know about the subject. This will help you pitch the material at the right level, and avoid being too technical or overly patronizing. An informal chat ahead of time, either with another teacher or with some of the students, may be useful here.

2. **Grab a lesson plan template**

   There are plenty of lesson plan templates available from Tes (a British trade magazine aimed at education professionals)—see www.tes.com.[4]

3. **Define the *O*, as in, *"Outcomes,"* part**

   Don't be too ambitious if you don't know the group. It will be most satisfying for you and your participants if you manage to conclude everything you set out in your plan.

4. **Create the *"Participatory Learning Activity"* part (second P)**

   Make this part modular—a good mix of lecturing, discussions, exercises and group work. Note that "participatory" refers to this not being a one-man show: everyone should be included. If you are ambitious, you can even turn the learning activity into a learning *experience*.

5. **Create the *"Bridging in"* part (B)**

   Now, try to be creative and come up with something that has the "wow factor" to pull people in and make them turn their full attention to you.

6. **Finish your preparations by adding the other parts**
   - **"Pre-assessment" (first P)**

     This could be a formal quiz, a free-form discussion or self-reflection, depending on the target group and the context. Be prepared to make some

---

[4] Tes Global Limited (2019). Here is an example: [https://www.tes.com/teaching-resource/ofsted-lesson-plan-template-11830234]

quick changes to your plan on the day, based on what you learn from the pre-assessment. Sometimes you will be surprised by the students knowing either much less or much more about the subject than you initially expected.

- **"Post-assessment" (third P)**

    Same thing here—quiz, discussion or reflection. If you have taught a skill, the "exam" could involve putting it to use.

- **"Summary" (S)**

    In its simplest form, this could be a recapitulation of the goals in the form of a slide with a bullet list. It can also be merged with the Post-assessment part, if this—for example—is a self-reflection.

    TIP: If you are running a series of lessons, you could also apply a technique used by skilled writers: let the end of your lesson (in the writers' case the end of the paragraph or chapter) point forward, and serve as a connection to or teaser for the next one.

7. **Set up a detailed time schedule**

Keep in mind that discussions and oral feedback can take a surprisingly long time. Another time trap is when the students return from their breaks; it is easy to lose up to five–ten minutes waiting for the stragglers.

8. **Run your lesson**

You will quickly notice that your thorough preparations make the execution part easy: the trail is marked out, and at every moment you know exactly where you are and what to expect. This way, time management also becomes easy.

9. **Make use of the feedback**

One of the main benefits with a closed-loop structure like BOPPPS is that you will have useful feedback material. First of all, the responses of the Post-assessment and the Summary parts help you get a grip of what you and the group actually achieved together; this feeling can be very satisfying if you usually finish your presentations without any feedback. But, more importantly, you will be able to follow your students' learning process and improve upcoming lessons on the basis of it.

10.7 Turning a Live Lecture into an Online Lesson

> **Purpose**: to give a lecture, a seminar or a training session to a group that is geographically distributed or that can't meet in one venue for some other reason.
> **Typical situations**: geographically distributed working conditions, such as online conferences and seminars.
> **Mistakes to avoid**: forgetting to adapt to the format; lack of interactivity; being boring and/or disorganized, losing the attention of the group.

Today, online meetings are an indispensable part of most workplaces, including academia. Of course, online courses and virtual seminars have existed for many years, but 2020–21 became a period where universities and their staff were forced to go through a huge transformation of mindsets, methods and culture.

Compared to pre-lockdown times, the general skills of a typical university lecturer in delivering online classes have significantly improved; at the very least, they have made the effort to acquire new know-how and to move beyond their old, in-person teaching habits. Some have succeeded by expanding their toolbox, finding hidden talents and learning new skills. Others have just turned boring and disorganized live lectures into boring and disorganized online lectures.

What are some easy wins to position yourself in the first category?

1. **Go through the schedule, the plan and the content of the live presentation**

   Ask yourself if it is realistic to keep the goal and the take-home messages of the live version; then make necessary modifications. Based on this tweaking, search the material for obvious things to leave out. Then try to take the leap from a live setting to the online setting—and as you do so, consider the following principles:
   - Your ability to interact directly will be restricted by the online setting. It will also reduce your capacity to engage the participants and control the intellectual atmosphere in the (virtual) room by your personality and presence.
   - Regular interaction is necessary in order to not lose the attention of the participants (even if you ask them to keep their video cameras on, your audience will be tempted to do things on the side while listening to you. You will be competing with social media, latest sports news, or a pile of laundry that needs folding[5]).

---

[5] Incidentally, keeping your hands occupied while listening to an online talk isn't necessarily a bad thing. Joanna realized that the only way of surviving a relentless onslaught of online meetings is by making beaded jewelry (though coloring, knitting or doing other crafts should work just

- Structure is always important, but even more so in an online setting. For example, make sure that you divide your slide deck into chapters—see Sect. 12.2.
- If possible, try to make the online lesson shorter than the live equivalent. Perhaps you can replace parts of your presentations with group discussions and exercises where the participants disconnect for a while?

2. **Based on the modified lesson plan, re-design your slide deck**

   This step is basically about adding and subtracting slides (designed in the same style as before).

3. **Introduce interactivity, further redesigning your slide deck**

   Here are some of your possibilities:
   - As an introduction, run a quiz or survey, using an interactive service like Mentimeter or Kazoo.
   - To maintain attention and motivation, run a short exercise, quiz or survey every 10–15 min.
   - Use breakout rooms for group discussions (remember to leave enough time for the logistics as well as for moments of confusion and small hiccups—time really flies when you're running exercises of this type![6]).
   - To deepen learning, assign larger tasks where the participants disconnect for a while, to work either alone or in groups. Here, the students can use cooperative online platforms like Google Docs or Dropbox Paper to present their assignments to you and the rest of the group.
   - Wrap up with a Q&A, welcoming questions both through audio and the chat.
   - To get feedback from the group, run a quiz, survey, or—even better—a self-reflection or review. This often proves useful as it may be hard to get informal feedback from the group in the virtual setting. If you skip this part, you may feel like you have been navigating blindly when the group disconnects, leaving you on your own.

4. **Make sure your audio and video quality is ok**

   Most quality laptops have adequate video cameras, though remember that decent audio quality is more important than high-resolution video and studio lighting. A headset or external microphone is strongly recommended.

   Don't study the video quality on your own unit. Instead, see what it looks like when you connect from another computer. You will probably notice that the quality of the video signal from your unit gets heavily degraded during

---

as well). Doing so helps to resist the temptation of sneakily checking emails—and lets your brain focus on the lecture. Plus, you are left not only with new knowledge, but also with a fabulous pair of earrings.

[6] Also, please bear in mind that the feasibility of it will depend on participant numbers and their initial engagement.

transmission. In an interactive video setting, "good enough" may therefore be a wiser ambition than "professional YouTuber quality."

Tips for your set-up

Set up good lighting

Make sure the main source of light is in front of you, to light up your face—though not shining directly at you (this will blind you!). Light from behind will mean that the audience only sees your silhouette, while lighting from above produces an unhealthy-looking set of shadows. A simple desk lamp can suffice—play around with the placement.

Eliminate distractions

Minimize distractions for both you and your audience. Run the workshop in a quiet space where you are unlikely to be interrupted (this means your housemate will have to open the door when delivery drivers come knocking). Turn off notifications on all devices, including your phone and computer—as well as nearby appliances, for instance in your kitchen.

Have a think about what your audience can see behind you. The background should be tidy, and ideally not too busy. Avoid sitting in front of a bookcase with a glass front—it will reflect your screen and provide further distraction.

Think about your posture and positioning

In-person speeches, bar panel discussions, are normally done standing up. However, for an online presentation, you will probably be sitting down—unless you have a standing desk (which we recommend anyway, for all sorts of health-related reasons). Make sure you are sitting in a comfortable chair with sturdy lower back support; a tip for good posture is the "BBC": Bum at the Back of the Chair (which results in a supported, straightened back, and not slouching).

On camera, your body language is, for natural reasons, limited. Don't sit too close to the camera so that only your face is showing; a tight crop showing just your face reduces your interaction even further. Instead, ensure that your arms and shoulders can be seen. With time, you will learn to wave and gesture so that your (restricted) body language is on camera. The good news is that your smile definitely comes through in the online format.

Maintain eye contact

Don't look at your own video image too much. This may be very tempting, but to address your audience as professionally as a TV host, learn to treat the green dot of your video camera as the face of an interested and engaged person. This may feel weird to many people—we speculate that it is even weirder for people with empathetic talent. Some speakers actually put a little image of a human face close to the video camera dot as a reminder.

Ideally, have a two-screen set-up; this way you can share one screen, and maintain any contact with the audience via the other one. This also helps if you want the participants to use the chat function for discussions.

5. **Practice and stress-test the system**

Before your first online lessons, there are no excuses: you absolutely must try out the setup, even if you are confident with the subject. The easiest way to do this is to run through a part of your lesson with a trusted colleague, friend or family member. You don't have to cover everything—focus on getting comfortable with the online format, and discover what tends to go wrong (a lot will, at least before you gain more experience).

While doing this, you should thoroughly examine the options of the software you use. Make all your blunders now, so you're fully prepared for the actual students. To get to know the system, run a few simulated online quizzes and exercises with your family or friends. Try word clouds and different kinds of graphs in Mentimeter, or let them answer questions in Google Docs that you will comment on—and so on.

6. **Give your presentation and keep learning**

Start your session by setting clear housekeeping rules. For larger audiences, ask participants to silence their microphones, while smaller groups can mute only when needed (for example, when coughing). Be prepared to silence the microphone of an individual participant if necessary, for example, if someone answers their phone during the session. Also, clearly define how questions will be managed—is this done during the session or at the end, and should they be spoken or typed? Most sharing platforms allow participants to raise their hand virtually, and to add comments and questions in the chat. You can also use an external solution for managing questions, like Slido, where participants can upvote the questions from others that they find interesting.

Since online sessions leave less room for body language and movement, try to compensate for this limitation with a more dynamic voice and facial expressions. Instead of pointing with your gestures, direct the audience's attention with built-in tools like virtual laser pointers available through screen-sharing.

When you meet your audience, the lack of human response—compared with an in-person lecture—can feel isolating. What's worse, talking into an online void created when the audience members don't turn their cameras on is extremely draining; you will also struggle to gauge engagement. Combat this by amping up your energy levels for the talk; make sure that you've had enough sleep beforehand, too. Perhaps, if you can, consider co-lecturing with a colleague—in this way, you can feed off each other's energy. Ultimately, trust your preparation and deliver with confidence, even when feedback feels minimal.

> **Further resources**
> Lots of incredibly useful video tutorials on how to set up online talks can be found on Jamie Gallagher's YouTube channel: https://www.youtube.com/@JamieBGall[7]

10.8 Giving a Media Interview

> **Purpose**: to explain or provide an expert opinion on a science topic, reaching a wide, non-specialist audience.
> **Typical situations**: interviews and talks for TV, radio or podcasts.
> **Mistakes to avoid**: being unprepared, boring or condescending; using too much jargon; persistently rejecting invitations.

Despite the popularity of social media and grassroot blogs and podcasts, traditional media still serve as the gatekeepers of what reaches the public in the form of news, ideas and trends. So, if you're approached by a journalist asking for a comment, or indeed for an interview with you in the starring role, how should you make the most of it?

Getting a visit by a reporter is, in most cases, a win-win situation. On the one hand, you or your research is interesting enough to warrant media attention (hopefully for the right reasons)—something that can lead to cooperation, funding, and future glory. On the other hand, the news outlet gets fresh and interesting content to present to their audience.

While there are many types of media and many ways of interacting with them, we would like to provide some advice applicable to most situations, before going into the specifics of interviewing for TV or radio.

Before you agree:

1. **Research the medium and the journalist**

    Listen, watch and read material from the medium in question, and focus on pieces produced by the journalist who approached you. Reflect on their mission, their tone of voice, how they approach the audience and what complexity level they have chosen for describing the world. In addition, get acquainted with their format and try to find out how your contribution will be presented.

    Figure out what is the top line of this interview—for instance, "overcoming a negative," "explaining a situation," and so on. This will depend on the program—is it edgy, intelligent, or newsy?

---
[7] Gallagher (2018).

2. **Be responsive**

   Journalists operate on very tight deadlines. Whether you can make the interview or not, let them know soon. Responding after a couple of days is usually useless, unless the journalist explicitly stated that there is no fixed deadline for a particular piece.

3. **Be available; don't worry about not being enough of an expert**

   Many science people spend their lives battling imposter syndrome—the feeling they are not good enough to belong in academia. This attitude, while unhelpful in general, results in turning down interview opportunities for which they'd be perfectly qualified. Due to societal norms and pressures (admittedly, not spread evenly across the globe), women are especially prone to doubting their own capabilities—which leads to a general dearth of female voices in the media. Journalists often say that when they approach a woman for an expert comment, they usually hear "what is the topic, am I competent enough, am I the best expert?", whereas a man will simply ask "when is the interview?".

   Our advice to this is: stop it. Just say yes. You know more about your subject than the majority of the population—and even if you don't have a Nobel Prize in your discipline (yet!), perhaps the Nobel Prize winner isn't particularly good at explaining science to the general public. While you are umming and erring, someone else (not necessarily better qualified than you, just more confident!) will happily take your spot.

   If you are researching a particularly controversial topic, you might want to run your media appearance past your PI[8] and your institution's communications team, and get their take on any topics that you should or should not address in the interview. The comms team might additionally provide you with extra guidelines specific to the outlet you will be talking to.

4. **Ask questions**

   Before going into the studio, find out about…

   …the topic:
   - What's the context; why is this topic chosen?
   - What questions will be asked?

   …the format:
   - Who will be there, who are the other guests?
   - Who is the target audience?
   - What will the final format of the interview be (will it be aired in full, or edited), when and where will it be screened?
   - Is it pre-recorded or live?

---

[8] Principal Investigator (PI) is the lead researcher responsible for a study.

...the logistics:
- Where is the recording taking place?
- How much time is required, when should you arrive?
- Will there be a make-up artist?
- How big will the shot be—entire body, just the portrait...? This will dictate what you should wear.
- Will you be compensated for travel?

Once you agree:

5. **Figure out your agenda**

You might think that the interview happens only in real time, when the journalist asks questions and you answer. This is not so: you must always go into the interview with a clear idea of what you want to say. Make this media experience work for you—what would you like the world to know?

Think about your key messages. Prepare a maximum of three short statements, 7–8 s each; they must be clear and consistent (for instance: "there can be no new drug developments without animal research"). Write them down for yourself to practice—they will act as your security blanket when you are nervous, because you would have rehearsed them in advance.

6. **Prepare your SPIN**

Summarize: what are your main points? This is why you have your key messages ready and rehearsed.

Prioritize: say your most important arguments first, since you might not have enough time to expand on all of them.

Illustrate: use examples and metaphors to make your case.

eNcapsulate: or put it all in a **Nutshell**. If you only have a minute to speak, how would you make your point? And if you only have ten seconds? This is the "give us a quote" moment of the interview, the soundbite that will be picked up and replayed in the teasers for the full interview.

On the day:

7. **Look your best**

If you are being interviewed for a radio or podcast, ask whether pictures or video will be taken (some programs do that). If not—it's one less thing to think about.

For video interviews, consider the following:
- **Clothing**. Avoid wearing busy patterns (they flicker on camera) and green outfits (you are risking the greenscreen effect and being a floating head). Find out what the site or studio is like, for instance, where will you sit? Will there be a table in front of you, or will you be sitting on a couch? If

the latter, avoid wearing a short dress for your own comfort—or if you do, make sure your underwear matches the color of your skirt.

Always wear clean and neat shoes—dirty ones draw unnecessary attention. If you are going to travel to the studio in the rain, bring a spare pair.

- **Make-up**. Will there be a make-up artist? If so, don't wear your own make-up beforehand. If not, wear make-up, bring your cosmetics and a mirror with you, and touch up when necessary. Yes, this applies even if you don't normally wear make-up, since studio lamps will make you look sweaty, shiny and washed-out. At the very least, invest in some translucent powder and a foundation that matches your skin tone.

- **Accessories**. Avoid wearing anything noisy, such as jingly earrings or bracelets. Shiny metallic accessories—as well as glasses—may reflect LED lights and screens, and be distracting for the audience. If you can, wear contact lenses instead of glasses.

8. **Go in and talk**

Interviews are like a tennis game. The journalist starts the serve with a question; you intercept it, and then point it in the direction you want it to go. The journalist picks up on an interesting part of your answer, and returns another question. You intercept it, and hit it towards your key messages—and so on, and so forth. If you give short, one word answers, the ball drops; if you go completely off-topic, the ball is out. Try to stay in the game, but keep it interesting. However, if you are asked about something that's not your area, it is acceptable to say "I don't have a view on this, can we move on?", as long as you don't overuse that statement. There's no point going down an unknown rabbit hole and wasting everyone's time by pretending you know things you don't.

Speak in a language that is understandable for everyone—using jargon alienates your audience; if you use the people's language, you will be one of them. Try not to create a divide between you and the viewers—relate to them, be a spokesperson for the people rather than a distant scientist in an ivory tower. When you say "we," refer to everyone, not just researchers. Never, ever be condescending.

Use short sentences, and have a start statement (especially if you've been asked to comment on a situation); journalists may use this clip as a soundbite. Short, sweet and focused is better than long, exhausting and tedious.

Ask whether the interviewer will appear in the final cut. If not, will you need to include the question within your answer, or will it be indicated otherwise, for instance through a caption or an edited recording? Ask when you're being recorded; be cautious of anything you say once you have been mic-ed up. Think about the persona you want to exude—enter that role before you get on stage, and let go of it only after you leave the studio.

When you are doing TV interviews, look at the interviewer, not the camera, unless you have been specifically asked to do the latter. Studio recordings differ from public speaking, because your movements are constrained by the work of the camera operator and the width of the shot. Try to stay as still as possible, and avoid looking from side to side—you're looking into a world the audience can't see, which makes you appear sketchy. Keep a natural facial expression while listening to questions.

How to deal with "trip up journalism"

Depending on the host, media outlet and topic, an interview can be supportive or confrontational. The legendary American journalist Mike Wallace said that good journalism is all about light and heat, where "light is the truth, heat is drama."[9] Some programs have a tendency to focus more on the "heat" than "light"—they thrive on spats and conflict, and specifically choose guests that are likely to spend more time arguing than having a meaningful discussion. While this is not very common, you might find yourself on the receiving end of controversial statements or questions. So what can you do if you suspect that a journalist is waiting to trip you up?

First of all, be polite and calm, and DON'T give in to the argument. Instead:

- Bat it back to the journalist, for instance with "I don't know why you are questioning me on that."
- Express sympathy, switch the perspective around, take the sting away from the question.
- Underline shared values.
- Deflect, come back to solid information (your key messages).

If you are being served an answered question (such as "You'd agree that…"), try answering with:

- "On the contrary…"
- "That's not the way I'd look at it."
- "I respect the right to varied opinions, though I don't share X's values."

Additionally, if your topic is in any way current, prepare by reading the latest news—this will help you avoid surprise questions.

Finally, if you have a controversial idea, have a think about whether it's really worth sharing.

9. **Build a relationship**

   After the interview is done, thank the team and try to find out when the material will be aired. Once you leave, evaluate your experience. Was it good or bad?

---

[9] Wallace and Knobel (2010).

What would you change? Is it worth doing again? If you were pleased with the result, maintain a relationship with the journalist who interviewed you (see Sect. 13.5); you may be able to pitch more topics to them in the future.

10.9 Giving an Official Speech or Informal Toast

> **Purpose**: to inspire a group to strive towards a common goal, or increase the sense of belonging to a social group.
> **Typical situations**: inauguration and closing of meetings; celebrations and special occasions.
> **Mistakes to avoid**: making the speech too generic; not being authentic; not addressing "elephants in the room."

There is a significant difference between a speech that wastes everyone's time and a speech that enters the audience's hearts: the latter is (1) personal, (2) honest, and (3) brief. We have all suffered through formal talks given by university officials, committee chairpersons and city mayors where every word was predictable, unimaginative and formal—creating a feeling of alienation, as well as an almost palpable impatience among those who are forced, by sociocultural norms, to listen.

The fact that people keep giving boring talks is disheartening, considering that the situations they give them in are often golden opportunities to ignite a spark in the listeners. You may motivate the audience to follow you *per aspera, ad astra*, 'through hardships to the stars.' Or you can simply give them something positive to remember forever—like a touching address to the students with whom you've shared thick and thin, or a heart-warming speech delivered at a funeral.

When you're given the formal task of delivering a speech, you should always adapt it to the social context. Consider the following factors:

- venue at which you are speaking;
- tone of the event (formal, relaxed, somber…?);
- context—including whether anyone (who?) is speaking before or after you;
- the organization or community that you are addressing, and what they are expecting to hear;
- the language you will—or should—be using;
- the message you want to leave your audience with—this is both explicit (your words) and implicit (what they feel, understand and remember from the speech).

All of the above need to fit together. For instance, in an after-dinner speech, when your audience is relaxed, full, and focused on digestion, they expect entertainment instead of, say, a deep intellectual challenge. But a sports team ahead

10.9 Giving an Official Speech or Informal Toast

of a game expects motivation. A research group after completing a project—acknowledgment. A grieving crowd at a eulogy—compassion. And so on, and so forth.

Below is a protocol for a speech prepared in advance. The authors of this book are the first to salute you if you decide to perform an *ad lib* speech at a festive event; the boldness of your initiative and your good spirit is much more important than perfectly executed details.

1. **Analyze the situation**

   What is the purpose of the assembly? What are the goals of the organization, social system, or shared-value platform to which you and your audience belong? Why are you there, and what do you want to achieve together?

2. **Compose the speech**

   You need three parts:
   - An intro, conveying the spirit of your speech. A quote or an anecdote is a safe choice.
   - A brief review of what you have achieved together, what this moment in time signifies, and what role individuals have played.
   - A call to action, a message of hope, an optimistic look into the future, establishing the feeling of "We're great!".

   The central aspect of the speech isn't its factual content, but rather the sentiment that "we are connected, and together we represent something grand and important."

   Finally, a note on brevity. When pondering the length of your speech, always think of the reaction you want to leave your audience with—is it "OMG never again!" or "More, please!"?

3. **Practice**

   It is a good idea to rehearse the entire speech before the occasion. Consider practicing both alone and in front of someone; and if you feel that it is too boring to rehearse your speech twice, you should probably cut down the length. Don't feel any pressure to learn it by heart; it is widely accepted that public speakers use notes on occasions like these.

4. **Execute the speech**

   Slow down the pace, take pauses, seek eye contact, and enjoy the emotional response.

   Example: A PI giving a toast at an end-of-the-year party.

   I recently read a quote by the great chemist Jöns Jacob Berzelius, who—as we all know—worked here at Uppsala University. This is what he said:

"The habit of an opinion often leads to the complete conviction of its truth, it hides the weaker parts of it, and makes us incapable of accepting the proofs against it."

I found this fascinating. More than 150 years ago, he warned us about the danger of cognitive dissonance—that we cling to our beliefs when we should actually keep an open mind. But, as your PI, I believe that cognitive dissonance is not a problem in our group. We've been challenging our colleagues all over the world with our three papers about cold catalysis.

On a lighter note, my wife claims I've been a bit obnoxious ever since we published our article in Nature. It is an observation that's hard for me to accept. So perhaps, on some level, I might be suffering from a case of cognitive dissonance after all?

This past year has been remarkable in many ways. Not only have we taken the lead in our research field, but we also have two freshly-minted doctors in our group. Sara and Zoltan: we hope you are as proud as we are of your amazing defenses and Ph.D. theses. Your findings and hard work have definitely injected new energy into the team. We old farts will do our best to keep up with you, but please, go easy on us.

Before I conclude, I would also like to praise our research assistant Domnika and our lab technicians Stefan, Hugo, and Sven. Without your hands-on approach and expertise, we wouldn't have been able to overcome some of the obstacles we've faced along the way.

Please, let's raise a toast—to our young talents, to our team spirit, and to the hard work we've put into our research! Let's make this new year one to remember!

Acknowledgements Thank you to Natalia Osica and Peter Ström for contributing to this chapter.

References

Gallagher, J. (2018). *Jamie Gallagher*. YouTube. Retrieved July 5, 2020, from https://www.youtube.com/@JamieBGall

Gallo, C. (2010). *Presentation secrets of Steve Jobs: How to be insanely great in front of any audience*. McGraw-Hill Education.

Strong, E. K. (1925). *The psychology of selling and advertising*. McGraw-Hill.

Tes Global Limited. (2019). *Tes—education jobs, teaching resources, magazine & forums*. Tes. Retrieved July 1, 2020, from https://www.tes.com/

Wallace, M., & Knobel, B. (2010). *Heat and light: Advice for the next generation of journalists*. Crown.

Instructions for Different Writing Tasks

11.1 Introduction

The American writer Gene Fowler (1890–1960) once claimed that "Writing is easy. All you do is stare at a blank sheet of paper until drops of blood form on your forehead."[1] We can all identify with this feeling, but—as we showed in Sect. 8.4.5—it doesn't have to be that bad if you follow some kind of method. In this chapter we present further writing tips for a range of tasks.

11.2 Articles for Peer-Reviewed, Scientific Journals

> **Purpose**: to disseminate your findings to an audience of peers; the most common academic "currency" for promotions.
> **Typical situations**: writing articles for specialist scientific journals.
> **Mistakes to avoid**: overly complicated language, lack of clear focus.

Every year, a myriad of original articles are published in a legion of scholarly journals (see Table 11.1). However, not all published articles are research-oriented, so it is important to know the differences between them.

---

[1] Weinberg (2006).

Table 11.1 A selection of some of the most common types of research articles

Types of Research Articles (adapted from the publisher Taylor & Francis)[2] https://authorservices.taylorandfrancis.com/publishing-your-research/writing-your-paper/different-types-of-research-articles/	
Book review	An opinion article of a published scholarly book.
(Medical) case report aka Clinical case study	A short report of a single patient's case with their symptoms, diagnosis, treatment, and follow-up. Requires a submission of consent forms from all human participants.
Clinical study report	An in-depth article with the methods and results of a clinical trial. Requires protocols to be accessible in a public registry to receive a clinical trial number (CTN).
Commentaries & letters to editors	Commentaries: responses to articles recently published; commentators can be invited to respond to a specific article. Letters to editors: short comments on current hot topics related to the field of the journal from readers.
Conference materials	Peer-reviewed, citable publication supplements in the journal that are presented at a conference as a record for scientific research. Example types: • Poster extracts • Conference abstracts • Presentation extracts
Data notes	Peer-reviewed concise articles of data held in a repository without analysis. Includes links of research articles that assists in increasing data impact, and promotes the reuse of research data.
Datasets	Any recorded forms of data from research deposited in a repository. Usually has a permanent identifier like a Digital Object Identifier (DOI).
Letters & short reports	Short articles of fresh data devoid of experimental details published to update a quickly advancing field.
Posters & slides	Academic research posters displaying experimental details and important results.

(continued)

---

[2] Taylor & Francis (2020).

(continued)

Registered reports	A report, where in the first stage of assessment, study protocols are submitted and assessed before data collection, and in the second stage, the full study is reviewed as an original research article.
Research article *aka* Original article *aka* Article	Classic IMRaD (Introduction, Methods, Results and Discussion) articles.
Review article	An analytical and interpretive review of already published research articles that find issues with the experimental data and offer suggestions for future research.
Software tool article	A description of the rationale behind a software tool development that includes the code, examples of dataset inputs and outputs, as well as how the output data should be translated. Usually written in open-access programming languages.

The most common type of journal article is the research article—it is the bread and butter of most scientists. Research articles usually follow the IMRaD (or IMRAD) structure: Introduction, Methods, Results and Discussion (Table 11.2).

Table 11.2 Unpacking the IMRaD format

Part	What is included…
Introduction	The broader context and specific aims of the study (research question, hypotheses).
Materials and Methods	How did you obtain data to answer your question (test the hypotheses).
Results	What data were obtained and what did they show (figures, trends, significance).
Discussion (& Conclusions)	How your data and analysis answer the question and what it means for the broader field; what are the next steps.

One of us (Olle) hates the format, because it's clunky, predictable and unimaginative. Another one of us (Joanna) argues that the point of scientific papers is not to be read as oeuvres of literature—although some do make excellent reading!—but to be able to quickly find what you are looking for. It's practical, in the same way that a dictionary is practical.

What makes IMRaD so popular? There are two main reasons.

- **Functionality**: as mentioned above, this format makes it easy for the viewer to localize the content that they want to study. It also forces the author to include all critical information necessary to advance the work.
- **Chronology**: the IMRaD format simulates an (idealized) timeline where the two parallel processes of science—the intellectual and the physical—align.

The structure of research papers fits into a double-funnel, or hourglass shape (see Fig. 11.1). The **introduction** is the broadest in scope: the overview of the topic gradually narrows to the specific research question of the study. It ends with your hypotheses. The **methods** section contains the useful details of procedure, sample sizes, equipment, as well as ethical permits. It ought to be written in a clear and detailed enough way to allow replicability. The **results** showcase the evidence obtained during your experiment. Figures help to illustrate your main points; while they are referred to in the text, a reader ought to be able to get the gist of one without having to read the rest of the paper. The **discussion** shows how your results answer your research question, and puts your findings in the wider context of scientific literature. This section also makes suggestions for future work, and points out any shortcomings of your study. It ends with a short **conclusion**.

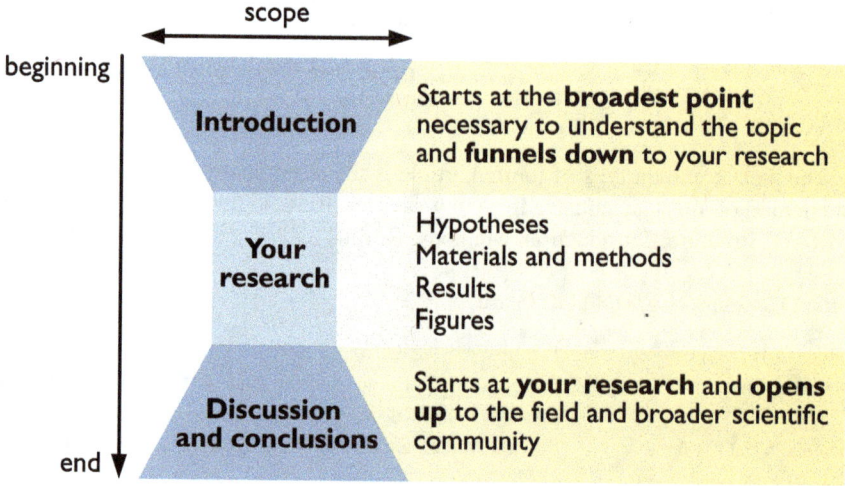

Fig. 11.1 The hourglass structure of a scientific paper

Apart from the main body of the text, you will need:

- **Keywords**: these help wider audiences find your article through search engines, in addition to what is already contained in the title and abstract.
- **An abstract**: the gist of the paper that includes all the key results. Remember, this is NOT a teaser-trailer ("Want to find out more? Read our paper!"), but rather a cheat sheet for anyone too busy or too lazy to read the entire thing. This is the part that is likely to be read by most people. Some journals will specifically ask for a bullet point format to ensure that the key results and findings are clear.
- **Acknowledgements**: an opportunity to thank anyone who contributed to the paper (intellectually, logistically, financially), but is not listed as an author. Be brief and specific here, and make sure you include all your funders.

- **References**: these should be current (aim for work not older than ten years, unless there is a good reason to do otherwise) and relevant to your topic. Format them in the style required by the journal; a reference manager like Zotero or EndNote will save you time and grief.

> **How other scientists will read your article**
>
> The most poorly hidden secret of science is this: **no one ever reads scientific articles the way they are presented by the journals**. When going through an article for the first time, no one starts reading from page one, no one goes through the columns of text in the proper order, no one studies the figures and the captions as they are mentioned in the text.
>
> Instead, since science people always look for time-saving shortcuts, they find their own paths through the article and visit the places where the useful information is found. Here's a typical route choice:
>
> *"I start by having a look at the Title, perhaps I also skim read the Abstract. Is it interesting?*
>
> *Then I have a look at the main figures and their captions. After that, I skim read the Conclusions/Discussion part.*
>
> *If it seems worthwhile, I have a look at the Introduction and the Methods.*
>
> *If the article is particularly important and interesting, I may study it further, perhaps even several times. Here, I am mostly interested in scrutinizing the best parts in detail. But I may also finally read it from the start to the end to get the whole picture."*
>
> It is worth bearing the above in mind when writing your paper.

Before we go any further, the authors want to make a quick note—in this section, we are not trying to provide a specific recipe, but more of a general set of guiding principles, applicable across numerous fields and journal types, that incorporate some of the bits of knowledge from Part II.

Chances are you have already written numerous scientific papers, but should you require more extensive writing assistance, there exist many dedicated blogs, articles, and books that go into great depth on how to write each section of a research article. We encourage you to explore these resources and consult the guidelines and recommendations from your favorite learned society. For example, the Council of Science Editors (CSE) or the American Association for the Advancement of Science (AAAS) provide valuable tools and recommendations to support authors in crafting effective research papers.

By leveraging these resources, you can refine your skills and ensure your work meets the standards of your discipline while effectively communicating your findings to the scientific community.

Now, let's continue.

1. **Know your story**

   Begin by defining your narrative. This doesn't necessarily mean introducing Cinderella into your paper on ecotoxicology, but rather thinking about the main narrative point that you are trying to make, and how your evidence supports it. This is the storyline that links the research question to the conclusion; the *raison d'etre* for the hypothesis, experimental design, and figures.

   A good starting place is sketching your key figures—they will correspond to your hypotheses and answer your research question, forming the core of your key messages. At this stage, decide what you would like to leave your readers with. In fact, some researchers don't write a word until the paper's narrative can be explained in a couple sentences.

2. **Pick your journal, check its scope and requirements**

   You probably have a fair idea, just from doing your background reading, which journals publish articles that are interesting, relevant, and seminal to your field. Still, choosing a journal to submit a paper to depends on a range of factors, including its scope, impact factor, publication costs, open access status, and so on. In addition, each journal has a different focus and emphasis on the scale of the question you are addressing. This is important to know when framing and discussing your work—does it need to be more theoretical or practical? Small scale or large scale? You should be clear who your audience is, what they would find interesting and useful, and what their level of background knowledge is.

   Once you have made your choice, have a look at the journal's authors guidelines for format, style, and word limits. They may differ substantially between journals, so it is better to know in advance what is expected of you. Also, remember that less is more: word counts are always more restrictive than you imagine.

   If you are not too terrified of coding, write your article in LaTeX—this will make reformatting much easier if you end up opting for a different journal at a later stage. Thankfully, many journals now accept raw manuscripts and formatting is done by the publisher's production team after acceptance.

3. **Start writing**

   A good way to kickstart your writing is with the **methods** section. Methods will stay the same regardless of your paper's storyline: they need to cover the study system, experimental design, procedures, and approach to statistical analyses. This section is the simplest one to write: after all, you know exactly what you have done, all you need now is to put it in words. It's an easy way to gain writing momentum.

   Once the methods are done, revisit the three or four figures that you've prepared earlier. They will help shape the narrative of the paper, that is, your hypotheses, your evidence and the answer to your research question.

The **results** should directly address the hypotheses. In this section, present all figures, describe the trends and outcomes, and show statistical findings.

With this structure in place, draft the rest of the paper following the hourglass format: start broad, narrow to specifics, and broaden again in the discussion. Use clear topic sentences (see Sect. 8.3.1) before writing out your paragraphs in full—it allows you to keep focused and not waste time.

A research article is a compromise between precision (detailed, specialist language) and readability. Try to keep everything simple: use short sentences, avoid jargon and overly flowery language. Write in the first person ("we weighed the water voles") and refrain from using passive voice ("water voles were weighed")—the latter is impersonal and boring.

In theory, anyone ought to be able to follow the article (unfortunately, in practice most papers are incomprehensible to a non-specialist; this does NOT mean they are good). Do not obscure your key messages—instead, highlight them, back them with evidence and position them within the wider field.

Don't get bogged down with crafting beautiful sentences—kill your darlings (see Sect. 7.2). Make your writing purely utilitarian; focus on what you say, not how you say it. Perfectionism is your enemy—be content with "good enough"! The point at this early stage is to get the key text on paper so the others can get up to speed on what you did and found, and advise on what is missing and what should change. Don't polish anything before receiving feedback; chances are you will have to rewrite it anyway.

4. **Get feedback and input from coauthors**

Research is a collaborative process, as is writing. The positive aspects of that include the cumulative expertise of the authors, a support network where everyone wants to succeed, and a variety of connections, experiences and insights.

On the flip side, you (especially as a doctoral candidate or an early careers researcher) will often find yourself between a rock and a hard place, that is, between supervisors and bosses with more gravitas than you, and strong opinions that might not align with yours or those of your other coauthors. They will give you contradicting feedback and argue over seemingly insignificant details. There is no easy solution, apart from letting the big fish fight it out between themselves—but bear in mind that the rubble left after that clash of the titans won't necessarily be what you set out to write[3]—and that's before the reviewers butt in! Keep calm and come to terms with the fact that you will probably end up with a camel (which, as the saying goes, is a horse designed by committee).

---

[3] During her doctorate on American mink, Joanna was caught between two supervisors, one of whom liked to write papers that were as broad and widely applicable as possible. He insisted she write about "small bodied, shallow-diving, semi aquatic mammals," to which the other supervisor would respond, in exasperation, "it's a f... mink, just call it a mink!" If you look at Google Scholar, you will know which of them got the upper hand in that argument.

5. **Write the abstract**

   Many scientists consider this part really daunting—but it doesn't have to be. Remember those topic sentences that you've been placing at the beginning of each paragraph? Take two most relevant ones from each of the IMRaD sections and glue them together; ensure that the Results ones contain key bits of evidence to answer your research question—et voila! You've got your abstract.

6. **Submit**

   Alongside your manuscript, you will need to write a short letter to the editor. In it, detail what your article is about, why it is interesting and how it is relevant to the journal. It's your opportunity to provide further justification and context to the work and explain why your paper is a good fit for the publication. Some editors may decide not to accept a paper not because of reviewers' comments, but because it does not fit the vision or mission of the journal—and for this reason alone they may decide not to even send it out for review. Therefore, know your audience and show that you do! Don't fall at the first hurdle after all your hard work.

   Once that's done,[4] let the waiting game begin.

7. **Deal with rejections and revisions...**

   Dust yourself off; fume quietly; be polite; implement changes; try again.

   If your manuscript is sent for review, it will, in all likelihood, come back to you covered in comments. While seeing so many comments may feel daunting and, at first sight, appear bad, in fact it is a sign that the reviewers took their time to ensure your work could be accepted.

   As you go through these comments, be unemotional. It may be difficult not to take all the criticism personally, but try to remove all emotion from the process and simply see it as a number of problems to solve. Be extremely polite and thoughtful, and you will improve your chances of acceptance.

   Address every point made. Accept all the easy minor editorial points first, then think about the moderate and major comments. Many of these will be justified points that you can and should address. Others may simply be suggestions that are useful, but arguably beyond the scope of the paper—here you need to justify it and form a rebuttal, a 'thanks but no thanks' response backed up by references. Discuss this strategy with your coauthors, and take a cue from the editor—what suggestions have they made for addressing the reviewers' comments?

---

[4] Although you'd be surprised at the additional detail you may need to provide, depending on the journal: highlights, popular summaries, short titles, options for graphical abstracts, photos and so on, all designed to help promote the paper's appeal.

In your final letter with the responses to reviewers, your aim is to make the editor's job as easy as possible: get them to approve the changes you made without having to resend the paper out for review again.

Lastly, accept the fact that the manuscript will have to be sent to multiple journals before finding a final home.

8. **...until you finally celebrate success!**

Once you publish, remember to make the pre-print accessible to a wider audience. Advertise your publication on social media and everywhere else—you deserve it.

Writing a short report

Of the many types of scientific publication articles listed earlier, none has as many aliases as does the short research paper. Also called **Brief, Rapid, or Short Communications/Reports, (Research) Notes, Correspondence,** or **Letters,** the names vary depending on the publication.

Short reports are usually written when an important research finding deserves to be shared quickly. These articles are tailored for a specific research community, particularly useful in fields that are highly competitive or time sensitive. Short reports are often followed up with a full research paper. One thing to note, however, is that short reports are not limited to researchers, and should therefore appeal to a broader audience.

Short reports differ between publications, so it is extremely beneficial and time saving to check journal-specific guidelines *before* drafting one. That said, there are features similar across the board:

1. **Brevity**: only the most noteworthy and arrestive findings are discussed; the length is often strictly limited so experimental details are not included.
2. **Speed**: some journals forgo or replace the peer-review process with an editorial review (other journals may simply publish more rapidly).

Due to the succinct nature of short reports, introductory information is not essential as you will be relying on the fact that your readers are already well versed in the background knowledge. Hence, your writing style must be accurate, and clearly communicate your findings and opinions.

11.3 Writing Effective (One-on-One) Emails

Purpose: all types of communication—private and professional, formal and informal, high and low[5]...
Typical situations: all types of professional and personal correspondence, job applications, scholarship applications, grant writing, et cetera.
Mistakes to avoid: Poor structure, missing information, lack of clarity about the topic and about what is expected from the receiver.

Please note that the text below is about one-on-one emails of the type that you write for a specific purpose and with a certain receiver in mind. As you are well aware, the bulk of the emails we are handling today are, unfortunately, of the automated, mass-delivered ilk. Let's leave them aside for now.[6]

First, a bit of history. The senior one of us, Olle, can still remember the first email he sent in Anno Domini 1989. Together with two colleagues, he managed to understand that the strange code he had received in a letter from a friend in the USA was some kind of glorified fax number. He recognized the weird "@" sign from his programming exercises in Pascal. The next day, everyone at his research department was excited when a reply actually arrived.

According to Statista, an international provider of market and consumer data, more than 300 billion emails are sent and received per day worldwide. Already in the late nineties, email became the backbone of professional communication in all types of workplaces. Academia actually served as a trailblazer of the email culture and was one of the first work environments where it became widely implemented.

Effective one-on-one emailing is a great demonstration of the goal-oriented communication approach we're advocating in this book. Anyone who writes a personal email obviously wants something from the receiver. The more well-structured, succinct and clear the text is, the greater the chance that this "something" is fulfilled (see Fig. 11.2).

---

[5] We know what you're thinking: "But surely anyone can write an email?" And yet the number of messages Joanna gets from her students that consist of "hi what's the assessment for this module," without any further details, prompted us to include this section.

[6] Something worth noting is how these promotional emails use principles we've discussed so far: the Inverted Pyramid, ethos/pathos/logos, AIDA, the elevator pitch, and so on.

11.3 Writing Effective (One-on-One) Emails

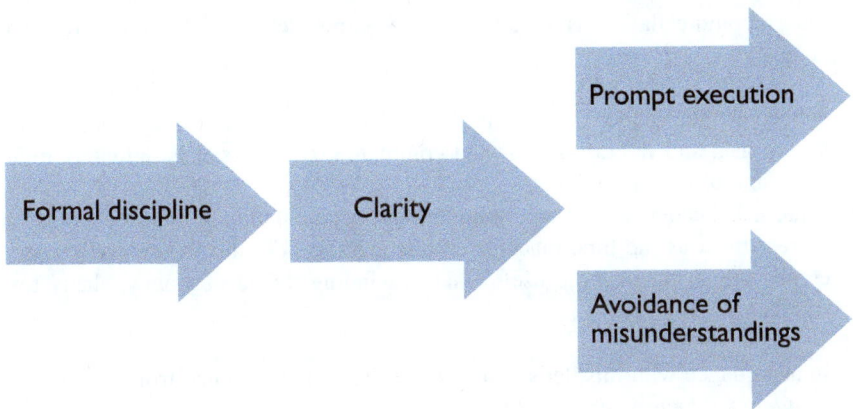

Fig. 11.2 With emails (like many other types of communication), formal discipline leads to clarity, which leads to prompt execution and lack of misunderstandings

Unfortunately, sloppiness still abounds. Every time we open our inboxes, it becomes apparent that there is much carelessness associated with email-writing. And the main reason for this is probably that different users look at the medium with very different eyes.

How about you, reader—which of the two metaphors below comes naturally to you?

- **The paper letter metaphor**

   According to this metaphor, email is an electronic heir to paper letters. This means that it is a medium for written language, based on writing and reading skills. Consequently, it is organized according to cultural and formal conventions (customary salutations, standard formatting, and so on).

- **The chat metaphor**

   According to this metaphor, email is a kind of glorified chat or messaging service. This makes it a hybrid medium—a written format which borrows heavily from spoken language, based on speaking and listening skills. Therefore, it is organized freely, to respond to shifting contexts and situations.

   When you study the result of these two mindsets, there is one important difference:
   - The paper letter metaphor assumes that the message is self-contained—no clarification is required.
   - The chat metaphor is an invitation to a conversation: "I'm here, ready to communicate about this—feel free to ask questions, and I'll fill you in."

Our recommendation is to stick to the **paper letter metaphor**[7]—for two reasons:

- It becomes the sender's responsibility to define exactly what they want to convey or ask, and to craft a clear and complete text. That seems definitely more fair—and also more polite.
- Email has become an extremely important documentation of our work; most of us rely on it as the time machine of our memory. For this, a few well-crafted emails are so much more useful than a winding thread of sloppy, chatty text fragments.

In accordance with this, let's first observe the correspondence from the view of the *sender*. As such, you want the receiver...

I. ... to understand who you are and what your message is about, to be able to assess its relevance and get an idea of its urgency.
II. ... to receive background information in a structured way.
III. ... to understand what is wanted from them, and how you would like to interact.

Let's then observe the correspondence from the view of the *receiver*. As such, you want to have replies to the following questions:

I. Who are you and what do you represent (organization, intent, values...)? Is this an authentic email, or is it some kind of phishing attempt?
II. Why have you approached me?
III. More precisely, what do you want from me or my organization, and how should I proceed to assist you, or at least give you a useful reply?

If you study the sender's and the receiver's views, you will see a 1–2–3 structure—which matches with the Orientation–Information–Action structure mentioned in Sect. 8.4.3.

1. **Define what you want from the receiver**

   As always, start by assessing the purpose of your communication activity.

2. **Start writing, using the Orientation, Information and Action model**

   Orientation

   This part includes the subject line, as well as the first part of the email body. Introduce WHO you are and WHY you are writing this message. Maintain a humble, empathetic and polite tone.

---

[7] At the very least, all new emails should start as this. If a chat ensues, you should be highlighting which parts of the text you are responding to.

11.3 Writing Effective (One-on-One) Emails

Information

Provide more context in the form of background details. Clarify the situation and the desired outcome, ensuring the receiver understands the full scope of the message.

Action

Conclude with statements or questions that prompt action from the receiver. If you want more than one thing to happen, list them in numbered format.

Example:

Subject: Borrowing a camera trap for a student-led scicomm event

Dear Professor Alfredsson,

I am Natalija Filipovic, project leader of the newly formed SciComm Club at the Biology Department. I bring a greeting from Dr. Jönsson in Environmental Sciences. I wonder if it would be possible to borrow one of the camera traps from your research group on Thursday?

The SciComm Club was formed last spring by students with an interest in scientific communication and science disseminations. We are running an event for high school students interested in wildlife, and would like to demonstrate the use of a range of surveying techniques. We already have a few live traps and tracking tunnels, but a camera trap would be really beneficial as well.

1. *Would it be possible to borrow a camera trap from your research group this Thursday, between 9 and 11 am?*
2. *If so, when would we be able to pick it up?*

Thank you very much in advance. If you have any questions, please reach out via email or contact me on the number below.

> *With best wishes,*
> *Natalija Filipovic*
> *Ph.D. candidate*
> *Environmental Sciences Division*
> *Pomeranian University*
> *+46 70 888 88 88*

3. **Things to consider**

- Make sure your whole name (as well as your affiliation) is mentioned somewhere in the email. It is, in fact, safest to set up and use an institutional email signature, which contains your name, role, department, university and phone number or physical address. If your name is difficult to pronounce, consider adding a "hear name" link, for instance from namedrop.io.

- Avoid abbreviations—being clear up front saves both parties' time.
- If you are writing to someone senior for the first time, it is better to err on the side of formality and address them by their title. What happens in the next email depends on the cultural context. In most Anglo Saxon countries, if "Professor Alfredsson" from the above example signs her response with "Anja," it is ok to use her first name going forward. In continental Europe, however, colleagues will address each other as Sir or Madam (or rather, Dr. and Prof., in academic circles) for years, especially if there is an age difference between them.
- If in doubt over what you should or should not say in an email, especially regarding sensitive matters, ask yourself whether you'd want anyone else to read what you wrote. Some topics are best discussed in person or on the phone.

The case of dates

Reading the Wikipedia article "Date format by country" brings a biblical story to mind: the one about building the Tower of Babel, and the global language confusion that followed. It is quite astonishing that the nations of the world have devised so many different ways to convey only three pieces of information: year, month and day.

Unfortunately, it is very hard to find regional or cultural patterns. Let's use the date 12th of February, 2023, as an example. In Sweden, it is written as 2023-02-12, like in South Korea and Russia. However, neighboring Norway and Finland write 23.02.12 and 12.2.2023 respectively. The greatest risk for misunderstandings comes from the custom of starting with the month, as is often done in the United States. The date above would thus be 2.12.2023. If a food package is stamped "To be consumed before 120223," it is hard for the consumer to make a decision.

What can we learn from this? It is often better to write out the full month name instead of using numbers, especially when working in a global context.

11.4 Grant Proposals

> **Purpose**: to obtain funding for research, development, teaching or other activities.
> **Typical situations**: any projects that require funding—academia, spin-offs, public engagement, NGOs…
> **Mistakes to avoid**: not dedicating enough time to the grant proposal, not asking for feedback from colleagues, not meeting the funders' objectives.

Writing a grant may seem like a monumental undertaking—and is indeed quite the time-consuming challenge no matter how many times you've done it before. What is more, the competition is usually fierce: for instance, in Australia, the National Health and Medical Research Council has funded only 20% of proposals received since 2017, while the US National Science Foundation (NSF) had an acceptance rate of 26% for 2021.[8] In fact, the molecular biologist Carol Greider had a grant rejected on the same day she received her Nobel Prize[9]! Still, grant-writing is one of the key skills in science, particularly since academic success and promotion often depend on the quality and quantity of grants received. Here, we give some tips on how to maximize your chances.

1. **Develop your idea**

   The fact that you're thinking about writing a grant means that you already have some idea for a contribution to your field. Now is the time to take a closer look at that idea—and make it stand out from the competition.

   According to most reviewers, the driving force of a good application is a robust hypothesis; after all, it is the foundation of your project. A strong hypothesis has **FINER** characteristics[10]: **Feasible, Interesting, Novel, Ethical** and **Relevant**—and while these are framed in the context of research, they easily translate into any project ideas you might have.
   - **Feasible**. Your hypothesis should be specific. It must be testable—within a reasonable budget, using existing technology.
   - **Interesting**. You—and hopefully the reviewer—should be excited by it. It should engage your audience by tying in the new with the old, bravely exploring new frontiers.
   - **Novel**. It should be fresh and bring something new to the table.

---

[8] Sohn (2019).
[9] In a graduation speech at Cold Spring Harbor Laboratory in New York in 2017, Greider said: "Even on the day when you win the Nobel prize, skeptics may question whether you really know what you're doing."
[10] Browner et al. (2022).

- **Ethical**. Your hypothesis should be answered through credible research conducted with the principles of confidentiality, informed consent, respect, integrity and so on.
- **Relevant**. It must be based on current scientific literature, have a strong rationale, and be important to the field. Even better if it can influence the field and lead to further study.

To assess what's been done and what's out there, a thorough literature review is mandatory. It'll also reveal if your idea and exploration question has already been done before, saving you from wasted effort and time.

2. **Pick the right funder**

Now that you have a solid idea, it is time to identify which grants will likely finance your type of project. Not all available grants may encapsulate the scope of your project, and not all project types will fit one grant program. Take your time to find the best match.

Once you have homed in on the funding call, get to know the grant application guidelines and instructions inside and out. Every grant has its own specific process, internal deadlines, guidelines, and formatting instructions that, if not followed, may get your application tossed from the pile. Unsure whether your project is eligible? Email or call the funding body to check. Beware of reducing your chances just because of formality issues.

If possible, study successful (and unsuccessful) proposals. Many funders—such as UK Research and Innovation, the US National Institutes of Health, or EU funding programs—will list successful grant winners, together with details about their projects, or indeed entire applications. More inspiration can be found via Open Grants, a website on which some 300 grants (accepted and rejected) are freely available to read.

3. **Start the process as soon as possible**

Let's stress this properly: START. AS. SOON. AS. POSSIBLE. The writing process will take a while; on top of that, you will want to leave enough time to obtain and implement feedback. Additionally, there will be internal processes within the university that you might not even be aware of; permissions you need to obtain, collaborators whose institutions require jumping through bureaucratic hoops—all of this takes more time than you think.

4. **Identify the necessary resources**

This will depend on your discipline, but by and large you can determine the common elements:
- How much money you require, and what you will spend it on;
- Who your collaborators will be;
- How long your project will likely take; and
- If you should conduct any preliminary studies.

5. **Choose your storyline**

   How you position your project—or what story you tell—should, as we know, be adapted for different audiences. Therefore, frame your argument and your narrative in a way that is relevant to the funding criteria rather than to you or your discipline; the two need to align.

   You are now entering the role of a salesperson, selling the funder a vision of success. As such, ensure your application contains four value propositions:
   - **Importance**. Does this research ask an important question?
   - **Success**. How likely is it that this research will answer the important questions?
   - **Value**. Is this project worth the resources it requires?
   - **Competence**. Does this applicant have the skills and knowledge to do this?

   The way you frame your research will determine how it is received, understood and engaged with. Build your narrative around change: emphasize what will be different as a result of your project. Be specific about your goals and how you will measure success. Remember that your work has to be achievable—overpromising risks sounding too grandiose and unrealistic.

   Writing a grant is no place to be humble. Justify why you, your team and your institution are best placed to undertake this work, should you be given the opportunity and resources. You are the experts, show what you bring to the table. Additionally, unique personalizing details will make your application memorable.

6. **Structure a project plan**

   Once you have set the tone and narrative of your proposal, it is time to get down to business and structure a project plan. While application formats differ, a project plan will generally be composed of two important parts: (1) **the background and significance** and (2) the **specific aims.**

   In short, the **background** is your chance to show how your research is relevant. Here, focus on how your proposal adds novelty, improves present knowledge, or solves a problem. Justify your claims with background information, which shows the reviewers you are an expert with a balanced viewpoint of the research area. Then you can point out any gaps you've identified in the field, proving yourself knowledgeable on the current state of your discipline. Avoid being wordy and beware of including unnecessary details.

   Then, in the **specific aims**, outline your project's objectives. Likely, you'll have several aims—make sure each is directly related to your overarching idea. Depending on the type of grant, each aim may require its own background and experimental method. State the alternatives and the reasons why you ruled them out. Be realistic: you should feasibly complete the project within a specified time frame. This will give the reviewers an idea of what they can expect, should your project be awarded.

   It is important for your research plan to meet all the requirements of the specific grant program. Keep in mind that although the plan includes all the technical details of your project, there are often strict page limitations. You must carefully consider what should and *shouldn't* be included.

7. **Think of the reviewers as you write**

   The grant reviewing process is deeply a human one, with human reviewers full of human incentives. These poor reviewers will often have to sort through hundreds of applications, so it is your duty to make the process as pain-free as possible. A proposal that is well-written, clear, unique and memorable stands a higher chance of getting funded.

   To make the reviewers' job easier:
   - Get them interested—position your work as novel and exciting, but also justified and achievable. Emphasize why the topic is important by providing a bigger picture explanation. Where possible, show how it benefits the end user: society.
   - Have a clear structure with a beginning, middle and end—don't waffle.
   - Don't overwhelm with details and jargon, keep the language as simple as possible.
   - Use cohesive devices for signposting.
   - Make your grammar, spelling and formatting impeccable.

   As you go through your proposal, check if it answers three fundamental questions that align with your value propositions:
   - Why you?
   - Why is this research important? (Or: so what?)
   - Why now?

   Finally, put extra effort into writing a strong summary section. Reviewers often go straight to it, because that's where the applicants succinctly, but convincingly, articulate their ideas and why they deserve funding. A well-written summary should inspire and reassure the reviewer.

8. **Edit your proposal—with feedback**

   Once you have put together your proposal, get feedback. There are a few ways of going about it.

   Firstly, engage your collaborators. Particularly if you're a little green about writing grants, facilitating an inter-department or even an inter-institutional collaboration may help lessen the burden of being the main writer. Sometimes, working jointly with a well-established laboratory can give you useful insight into how others write grants.

   Secondly, engage experienced colleagues or mentors. An important ingredient, so to speak, is to have access to people who can help guide and support you in your writing journey. If you don't have one yet, find a mentor who has had grant successes and knows about your field. Remember, however, that they, too, have projects and lab work—and even if they are generous with their time, your grant isn't necessarily their priority.

   Thirdly, engage external help—particularly if the grant is a high-stake one, and you have the funds to pay for the service. Hiring grant writing companies, professional grant writers or grant editors is more widespread in some parts of the world than in others.[11] They offer assistance with anything from language support, proofreading and minor edits, to funding searches, project management and writing the actual proposal. The costs will vary depending on how much help you need. Larger academic institutions may have either grants teams, or editors at the institute's writing center, who provide similar services free of charge—so check what is available to you first.

Whatever method you choose, make sure you leave yourself enough time to receive and implement feedback before submission, and send off the best possible version of your proposal. You have but one chance, after all.

Writing a winning grant proposal
Tips from Jessica Vargas, Grant Advisor, Innovation and Impact Center, TU Delft, Netherlands.

- Before writing your proposal, check the funding program and make sure your idea is in line with the aims and scope of the program. Ensure eligibility.
- Carefully read the guidelines and strictly follow the template when preparing your proposal.
- Tell a compelling story. As a grant writer, you aim to create a narrative that catches the reviewers' attention. Follow a logical structure and make sure your ideas are presented in a clear and concise manner. Be explicit about the message you want to convey.
- Align the budget to the project narrative. The budget is an opportunity to show the project is feasible and well-planned.

11.5 Press Releases

Purpose: to communicate with the media effectively and clearly; to develop a good working relationship with journalists.
Typical situations: promoting your group's research and providing material for newspapers, magazines, and websites; keeping other interested parties updated.
Mistakes to avoid: being too pushy and self-glorifying; not including the embargo date; not using a standard format; wordiness.

---

[11] Sarang and Joanna had a long and heated discussion over this recommendation!

Since their creation in 1906 by Ivy Lee, press releases have continued to be one of the most common methods to disseminate science to the public. Nowadays, press releases are tools for branding and credibility, reaching policy-makers, scientists, experts in related fields, and beyond. With such a wide reach, it is imperative to write in a manner that is accurate but concise. Here are some of the main goals of a press release:

- To bring attention to progress in science and technology.
- To advertise an institution's efforts.
- To highlight the results of scientific projects.
- To highlight the relevance or applicability of scientific findings.

Sending out a press release may have four different outcomes:

1. It is thrown in the bin and neglected. This is a very likely outcome, but persevere and keep'em coming!
2. It is read by a journalist who decides not to take further action. However, a seed of interest may be planted. Same thing here: persevere and keep'em coming!
3. It is rewritten and maybe even quoted by an editor and ends up in a section for short news pieces. Excellent! Persevere and keep' em coming!
4. The media outlet sends a reporter and perhaps even a photographer or camera person to your lab for a feature job. Congratulations—now read Sect. 10.8 to prepare for the interview.

It is possible that your institution has a **Communications Team** or a **Public Relations/Press Officer** who may draft the press release in collaboration with you or the principal investigator. The communications team will know the most suitable destinations for a given press release; they will also advise on the timings of certain pieces (you don't want your news to compete with something bigger or more trendy).

Your press officer will also be able to tell you what the media find attractive and relevant. While your research group may be very excited to announce the news of a huge grant, the press is more interested in the outcomes of the research that it funds. Similarly, not every single paper you publish will be news-worthy; reserve press releases for the most impactful studies. Trust the expertise of the press officer.

However, institutions may be tight on resources, so you may find yourself writing a press release on your own. Additionally, even if you work with a communications specialist, you still need to provide them with the bulk of your content—how else would they know what to write?

Bearing all the above points in mind, here is a recipe for structuring a press release (remember the Inverted Pyramid from Sect. 8.4.2 when you set out to write).

11.5 Press Releases

1. **Choose the embargo date**

   An embargo date is particularly important if your press release is based on a scientific publication. Press releases **must** be sent to journalists in advance of the scientific journal publication date and **should not** be published until the journal publishes your article. The **embargo date** tells the journalist when they can publish the article, but also ensures the timing of when the information is released. Without the embargo, data and analysis may be made public prematurely, which could influence the initial impact of the research paper. To prevent this, the date should be determined first, after discussions with your PI, collaborators, and the editor of the scientific journal in which your research paper is published.

   Note: an embargo date is usually applicable for those working in a professional or academic setting, and may not be necessary in other contexts.

2. **Add media and interview contacts**

   Add your contact details as well as the details of people who are willing to be interviewed about your research. Those who have agreed should set aside the day of release in order to respond quickly to journalists who may make an inquiry or ask for a secondary opinion after receiving the press release; they should also be available post-publication. To aid the interviewees with their preparation, try to send them a reminder several days in advance of the publication date.

3. **Create a title and subtitle**

   Warning: this part may take the longest!

   You may feel averse to writing a click-bait title. Yet, without an attention-grabbing header, it's likely that your press release may drown in the deluge of the perpetual shock and awe of pitches, alerts, and press releases journalists receive daily. Editors will judge whether to read the release in full based on the title alone.

   > **Step 1.** Put yourself in the journalist's shoes and ask yourself, *Why would I care about this press release?* The answer may be the main ingredient needed for the headline to capture the journalist's attention.
   >
   > **Step 2.** With the answer, keep in mind that in the world of social media soundbites, the title should be newsworthy. Make it short—try to keep it within ten words. It must be informative and easy to understand. Do not exaggerate! Be sure to include only relevant scientific details, but avoid using scientific jargon. Remember, the press release will reach a vast array of individuals, many of whom may not have the necessary background to understand the particular branch of science you practice. This includes the journalist to whom you are sending the article.

Step 3. Once the title is ready, use data and statistics in your subtitle to set a tone of credibility. Your subtitle should also be succinct, speaking directly with indispensable information.

4. **The first paragraph**

   Begin with the Five Ws: Who, What, When, Where and Why. The main findings and the significance of the impact should be summarized here. Write in an active voice and avoid scientific language. Don't overload information; focus on one or two key points.

5. **Second paragraph**

   Mention who was involved in the research and briefly explain the work process while further contextualizing the importance of the findings. Reiterate the key results.

6. **Subsequent paragraphs**

   Within the context of the research, and with their consent, add engaging quotes from the authors or other scientists familiar with the study. Include the applicability and relevance of your research, as well as the goals of the research group with regards to the project. Finish with "[ENDS]."

   It is important to never overhype or oversell the research. As scientists, we are responsible for the accuracy and integrity in our communication; we want to avoid situations where readers misunderstand the findings and form an incorrect belief about the topic.

7. **Who should be notified?**

   While you are writing a press release, keep the following people in the loop:
   - The authors of the research paper that you'd like to publicize.
   - Press officers of your respective institutions (even if you don't have one, your collaborators might—and they would usually be happy to assist).
   - The editors of the journal where you are publishing your research. Make them aware that a press release is being written; they will inform you of your article's publication data and URL (include these details in the press release).
   - Any potential interviewees—this includes the study team, but may also extend to, say, patients who are affected by the disease you are combatting. Make sure they have consented to give quotes or interviews.

   Once the release is ready, send it to your media contacts.

> **Effective Press Releases**
> *Three tips from Heather Buschman, Ph.D., University of California San Diego. Dr. Buschman is a press officer at UC San Diego Health and teaches a science writing course at UC San Diego Extension.*
>
> **Timeliness.** A press release is issued by your university's press office—contact them early when you have news or a paper accepted for publication. Releases are proactively distributed to members of the media, and are intended to pique reporters' interest so they can develop their own stories. To a reporter, a paper published yesterday is often considered "old news," so press releases should be developed ahead of time and issued while the news is fresh.
>
> **Impact.** Don't be offended if your press office declines to issue a press release on your news item. Even if they had the time, it would be detrimental for them to issue a release for every paper, grant and award. Members of the media receive hundreds of pitches each day. The more an institution sends, the more reporters get (understandably) annoyed, and true breakthroughs don't stand out. Press offices tend to prioritize the news items that are most likely to catch the media's eye—trust their expert judgment! Fortunately, most institutions also feature news via other channels, such as blogs, newsletters and social media.
>
> **Accessibility.** Press releases are usually intended for non-scientific audiences. Your press office team will appreciate your help in describing your work without getting too far down in the weeds. Avoid technical terms, acronyms, unnecessarily fancy words and long sentences. Instead, suggest metaphors, anecdotes, emotions and images to help engage readers and demonstrate that scientists are people too.

11.6 Writing for Mainstream Media

> **Purpose**: to convey the basic facts about any subject in a reader-focused, easily understood way.
> **Typical situations**: writing news articles, explainers, opinion pieces, press releases.
> **Mistakes to avoid**: not getting to the point immediately, using subjective or promotional language, using jargon.

Please note that writing for mainstream media is closely related to the press release, Sect. 11.5. We suggest reading these sections side by side.

There might come a time when, instead of focusing on your regular academic writing, you find yourself in the role of a popular science writer. This could be because a journalist asked you for an expert comment, an editor commissioned a book review or an explainer, or you simply feel the need to write a letter or an opinion piece. Even though the scenarios might differ somewhat, we'd like to provide some general guidance on writing for a non-specialist audience.

1. **If there is any: go through the content brief**

   If you have received an assignment from pros, their brief should at least contain the following:
   - A content summary or a working title,
   - An overview suggested by the editor,
   - A word count,
   - A deadline.

2. **Decide on an angle and a main message**

   Begin by brainstorming the messages that you would like to convey (see Sect. 8.3.2 for guidance). Then go for a cup of coffee and reflect for a while on your notes before you sit down to write the first draft. What is the story here?

   > **Finding the right angle**
   > A central concept in news reporting is the so-called **angle**—that is, the point, or the theme of the story. Ask yourself: "At this particular moment, what makes my text most compelling and relevant for our target group?" In most cases, there are multiple angles to a story; it is part of the reporter's skill set to continuously find angles that the readers appreciate.
   >
   > *Example:*
   > *In the summer of 2022, some places in northern Europe were visited by a roving walrus, a species that is rarely seen in this part of the world. Freya, as she soon was called by the public, caused some stir when she sun-bathed on small yachts of her liking. In the international stream of news pieces that Freya's travels generated, the following angles could be spotted:*
   >
   > - *Nature clashes with civilization ("wrecks boats").*
   > - *Roving beast spotted in different places (NL, UK, DK, NO, SE).*
   > - *Odd animal draws attention ("goes viral on the internet").*
   > - *Confused critter stressed by curious people ("zoologist concerned").*
   > - *Climate change causes odd behavior of walrus.*
   > - *Zoologist suggests shooting walrus.*[12]
   >
   > In science and tech journalism, some common and useful angles might be:
   > - human interest/humanitarian importance,

---

[12] Unfortunately, this advice was taken ad notam by the authorities. R.I.P. Freya.

- solution to a problem,
- old mystery solved,
- opportunities for new science,
- stirring controversy,
- innovation potential,
- economic impact,
- out of the ordinary.

When it comes to timing and relevance, media pieces are split into two categories: pegged and evergreen.

Pegged pieces are the ones that relate to a particular event, occasion or news. They are usually written with the purpose of understanding a current event or phenomenon better. They are divided into:
- **tight pegs**, or those closely linked to an event (such as elections, crises or recent news). Note that because everyone will be rushing in to comment, you will have a lot of competition—although academics often have interesting and original takes on a subject.
- **loose pegs**, which tend to have three to five years' relevance. This could, for instance, be "something about social media"—a piece that an editor can whip out if a topic happens to be trending. However, if the connection to a phenomenon or event is barely there, don't force it, just write it as an evergreen.

Evergreen pieces are those that transcend the news cycle and can run at any time. On the plus side, you won't be competing with others who want to cover the same topic; on the other hand, you are at the mercy of the editor's whim, since the piece could be published tomorrow or a year from now.

Decide whether you would like to position your piece as an evergreen or a peg; it is best to consult your editor in this matter.

3. **Write a placeholder headline and the lede (lead paragraph)**

The **headline** should tell the story as effectively as possible, preferably based on at least one active verb. Bear in mind that you are writing it mainly for your own benefit—it's a great exercise in figuring out what the main message should be—since most of the time the final title will be picked or written by the publishing team.

The **lede** (or *the lead paragraph*) is, according to Wikipedia, "the opening paragraph of an article, essay, book chapter, or other written work that summarizes its main ideas."[13] Let yourself be guided by the Five Ws (Sect. 4.4.3) here. If the medium you're writing for has very short ledes, there may not be room for all the basic facts; in that case, include them in the second paragraph.

---

[13] Wikipedia Contributors. (2019).

> **Examples of journalistic jargon**
> **"Bury the Lede"**
>
> Negligence to follow The Inverted Pyramid principle, instead burying the interesting stuff inside a bulk of irrelevant details. Science people: this is looking at you!
>
> **Byline**
>
> The writer's name at the top or the bottom of the article. Sometimes includes a photo and/or contact information.
>
> **Copy**
>
> Text produced by a journalist or text professional in the advertising industry (copywriter).
>
> **Sidebar**
>
> A shorter text that accompanies the main text, for example a biography, a brief interview, a fact box. See Sect. 8.4.1 about chunking.

4. **Now, with the headline and the lede as your guides, add more paragraphs to the text**

   At this stage, the headline/lead section serves as your main argument. Whatever you add to the text should clarify what you convey at the start.
   - The mid section, or the **nut graf** (from "nut paragraph") of the text normally fills in the blanks and brings more details, explaining the context of the story "in a nutshell."
   - Further sections give additional background and "nice-to-know" info.

   You will note that this procedure simplifies all the micro decisions that writing consists of: if something is relevant to the headline and the lede, then include it; if not, leave it out.

   Remember: during this process, keep checking the word count every ten minutes or so.

5. **Refine your text**

   The difference between popular and scholarly writing is the readers' patience. Reading a popular piece until the end is not mandatory—so think about what happens if people don't finish the whole text. What will they know if they stop reading after the lede? Or the nut graf?

   Consider how a piece will land: not just what you want to say, but what people are willing to hear (and how). Signal subtlety upfront, otherwise the readers might not get to it.

Avoid getting lost in nitty–gritty detail and confusing jargon. Never make the general reader feel stupid. Let your editor guide you—if they tell you something needs changing, do it; they know what they're talking about.

When you're done writing, put the text to one side for a while. Then go back to it and read it out loud to yourself. Happy with how it sounds? Send it off!

Tips for writing an opinion piece

Opinion pieces are a powerful way to stand up for evidence-based knowledge, bring important issues onto the public agenda, describe problems and propose solutions. When timed correctly, they can also influence decisions—including the launching or halting of political, judicial or administrative processes.

Here is our advice:

Act fast! In the media world, the window of opportunity closes very quickly. Do not wait until after the weekend. Instead, start researching, outlining, and calling potential co-writers already this afternoon. No matter how important you find the issue, newsrooms quickly move on to the next topic of conflict … and the next, and the next …

Sit down to read. Follow current discussions and find out the background and stance of those who have engaged so far. Call a trusted friend or—if applicable—your co-writer to discuss the subject, and try to identify messages, arguments, and the weak spots of your opponents.

Build a strong case. Before you start typing away, ensure you have:

- A strong and well-defined opinion on a core issue. Stay concerned, passionate, and focused.
- At least three factual arguments, supported by hard data, examples or quotes from reliable sources.
- A call to action that aligns with your stated position.

Follow a three beat structure. Present three key thoughts as soon as possible. These are:

A. What the readers know
B. What the author is talking about
C. Why does it matter that the readers learn what the author knows
      or: A **but** B, **therefore** C.[14]

---

[14] Another way of approaching it is:
 (A) Here's what we're talking about.
 (B) Here's the challenge.
 (C) Here's how I propose to solve it.

Get to the point—fast! Show which side you're on. Let the world know you are distressed, annoyed or worried by the current state of affairs. By the end of the second paragraph, the reader should be able to tell what your main message is.

Be brief—it is more effective than being wordy. So, 800 words or less is better than 900 words or more. Focus on, at most, one idea. Don't hesitate to go small, since there isn't much space. Know what to leave out without the detriment to the science argument.

Choose examples wisely. Pick just one example per point—you are not trying to build a case with evidence, but illustrating a concept.

Keep it personal. The use of "I" is helpful to emphasize the opinion or the salience of the perspective.

Use humor if it helps you make your point. But please avoid it if it makes you appear heartless and indifferent. Irony can be effective, but it can also cause confusion about your actual meaning.

For more advice on opinion pieces, follow Joseph Fridman's channel on YouTube: https://www.youtube.com/@Yossi842[15]

11.7 Writing an Interview for a Magazine, Newsletter or Blog

Purpose: to produce content for a blog; learn new stuff directly from the horse's mouth; extend professional networks.

Typical situations: maintaining a personal blog, contributing as a community member of a themed blog.

Mistakes to avoid: formulating questions to the interviewee which give irrelevant replies; making the text too wordy.

If you intend to write a blog interview, you're probably running your own personal blog, or contributing to a multi-author blog. Here's a dirty little secret about these kinds of simple email-based interviews: they are easy to write, and the content automatically gets fact-checked by an expert of the subject at hand. How come? *Well, because the interviewee writes most of the text.*

Because you put the burden of text-production on the interviewee, you have the responsibility to make their effort worthwhile:

---

[15] Fridman (2006).

- Make the process smooth, so they don't waste their time: be up-front about expectations, timelines, and processes. Administrate email correspondence carefully and stick to a time schedule.
- Don't ask too many interview questions. You will probably be surprised at how much text you get in return: much of it will be edited out. It is very frustrating for the interviewee to put effort into writing the text just to see it being deleted during your editing.
- Do your best to disseminate the text, reaching as many readers as possible from the target group. Maximizing reach is a key way to honor the effort you and your interviewee have put into this project.

1. **Pick an interesting person and do your homework**

    The initial research you do on your interviewee serves two purposes:
    - Primarily, it helps you ask compelling questions, resulting in richer replies and a more interesting read for your followers.
    - Secondly, it gives you credibility when you get in touch and ask for their time and effort.

    Strive to engage in whatever is important in the interviewee's life and mission. If they are scientists, go through their CV and study their most important publications. If they run a podcast, listen to a few episodes; if they're a writer, read their books.

2. **Send an initial email to the interviewee**

    This introductory email you send to the interviewee should contain the replies to the following questions (seen from their point of view):
    - Who are you?
    - How did you learn about me?
    - Why do you find my mission important?
    - What do you want from me?
    - OK, what do we do now?

    Example
    Subject: Interview for Funky Relativity blog?
    Hi Niels,
    I'm Albert E., physicist and writer. I run the blog Funky Relativity, where I cover things happening in the field and do interviews with influencers.
    Recently, I listened to your podcast Quantum Qupenhagen and was intrigued by your ideas of using matrices in quantum mechanics. That would definitely make the calculations a lot simpler! I am sure the readers of my blog would love to hear you develop your thoughts around this.
    So, would you be interested in doing an interview for Funky Relativity? If you like the idea, my suggestion is that (1) I send you some questions by email, (2) you reply to them and (3) I edit the material into a finished blog article. From my experience, this process usually takes about two weeks.

> *Of course, you will be able to have a look at the finished result before it is published. We will do our best to disseminate the interview through our networks.*
> *Looking forward to continuing this correspondence!*
> *Greetings from Vienna,*
> *Albert*

As you are likely not the first to find your interviewee interesting, don't be surprised or disappointed if you may have to wait for a response. After a week, follow up. After two weeks, send a kind reminder asking "Just wondering if you received my email?"; this time, you may use another channel, such as an alternative email address or LinkedIn.

3. **After approval, send your questions and get replies**

   The first rule regarding questions for email interviews is: be succinct! If you write a 10,000 character text (including spaces), very few blog visitors will read it till the end. So try to come up with 4–6 clever, thoughtful questions in your first round of correspondence—that will probably give you enough for a 5000–6000 character article. If needed, you can follow up with two or three more questions.

4. **Edit the material**

   If the interviewee is a talented, experienced, and engaged writer—as many influencers are!—you may just have to come up with a title, a lede (see Sect. 11.6) and do some formatting. If you are more ambitious, you can edit it more thoroughly; see examples in the box below.

Editing an interview

Example 1: Q&A interview—simply publishing questions and replies

Q: How come you have chosen to stay in your homeland as a scientist?

A: Denmark may be a small country, but Copenhagen is a great place for a physicist; I am surrounded by brilliant colleagues and we can really rely on the support from The Carlsberg Foundation.

Example 2: an edited interview

Niels B. has chosen to stay in his native Denmark to pursue his career as a scientist. Although it is a small country, the academic environment of the capital offers many possibilities:

"Copenhagen is a great place for a physicist," Niels says. "I am surrounded by brilliant colleagues and we can really rely on the support from The Carlsberg Foundation."

5. **Show the finished text to the interviewee, make the final changes and press Publish!**

   Always make sure that the interviewee is happy with the edits before the text goes live.

6. **Disseminate the article in social media**

   Ask the interviewee for all relevant social media handles, and put a lot of effort into the publicity. Remember, this is the only payment they are receiving, so make it worth their while. Nobody wants to spend time poring over a piece read by only three people.

Acknowledgements Thank you to Helen Cullen, Joseph Fridman, Ada Grabowska-Zhang, Natalia Osica, James Ryerson, Jessica Vargas and Richard Walters for contributing to this chapter.

References

Browner, W. S., Newman, T. B., Cummings, S. R., & Grady, D. G. (2022). *Designing clinical research*. Lippincott Williams & Wilkins.

Cold Spring Harbor Laboratory. (2017, May 9). *Nobel laureate Carol Greider on fighting doubt*. Facebook. Retrieved November 14, 2022, from https://www.facebook.com/watch/?v=1420793747941053

Fridman, J. (2006). *Joseph Fridman*. YouTube. Retrieved July 15, 2020, from https://www.youtube.com/channel/UCdT8JMidpWhTGmUGYtkKmqg

Sohn, E. (2019). Secrets to writing a winning grant. *Nature, 577*(7788), 133–135. https://doi.org/10.1038/d41586-019-03914-5

Taylor & Francis. (2020). *Different types of research articles*. Taylor & Francis: Author Services. Retrieved July 8, 2020, from https://authorservices.taylorandfrancis.com/publishing-your-research/writing-your-paper/different-types-of-research-articles/

Weinberg, G. M. (2006). *Weinberg on writing: The fieldstone method*. Dorset House.

Wikipedia Contributors. (2019, January 25). *Lead paragraph*. Wikipedia; Wikimedia Foundation. Retrieved July 9, 2020, from https://en.wikipedia.org/wiki/Lead_paragraph

Some Tips for Visual Communication   12

12.1   Intro

In Chap. 9, we introduced the concept of visual communication, offering some essential advice to "reluctant visual designers," as well as more detailed tips applicable to a variety of contexts. Here, we're taking things one step further, and jumping straight into specific design tasks.

Please note that our goal is to give you some starting points for self-studies and exploration. If you want to go further, we suggest taking some online courses in subjects like vector drawing, photography, raster graphics editing, and basic layout. There is also hands-on know-how to be found in our list of favorite handbooks (see the top ten list in Further Reading).

12.2   A Slide Deck for a Short Scientific Presentation

> **Purpose**: to summarize your research for different audiences; to disseminate results; get feedback and attract cooperation.
> **Typical situations**: conferences; group and faculty activities; teaching; industry interactions.
> **Mistakes to avoid**: poorly defined main messages; lack of structure; too much material/too many details; misguided designer ambitions.

Succinct presentations lasting 5–15 min are the bread and butter of scientific peer-to-peer communication. Without a doubt, it is a versatile format used in professional meetups, conferences, symposiums, workshops, and educational activities.

In this section, we'll guide you through planning, preparing, and producing a well-structured slide deck for presenting "typical" research projects from different scientific domains. We have settled for a 10-min format: it is enough to cover a standard research project, yet leaves time for 1–3 min of Q&A and discussion.

As our starting point, we're using the *"Outline of a Structured Scientific Talk"* by Matt Carter[1]—a 16-slide, carefully structured presentation template, appropriate for conferences or faculty seminars.

Note: In Chap. 7, you can learn much more about presentations in general, and how to plan, prepare and perform them.

1. **Define the goals of the presentation**

   In line with Chap. 6, we start by setting the goals. Here are some examples for this particular context:

   - **At a conference**: Disseminate recent results, initiate peer-to-peer feedback, establish the reputation of a research group, attract contacts.
     What to do: *Tell a well-crafted story without loose ends; make the presentation appealing for potential partners.*
   - **At a faculty seminar**: Keep colleagues informed about ongoing projects, solve practical and theoretical problems, initiate discussion.
     What to do: *Tell both sides of the story. On one hand, describe successful experiments and results; on the other, point out problems and address impediments. This is no time to be shy, since the collective intelligence and experience of the faculty are at your service to solve these issues.*

2. **Analyze the main messages, the target group, and the context**

   Revisit Sect. 6.2 for guidance on tailoring your core messages and considering the presentation context.

3. **Pick the visuals that tell your story**

   By this, we mean the visuals that dictate your narrative: your key figures, photos or diagrams; the ones that demonstrate success or prove your main point. These will shape the structure of your presentation, and should be chosen early in the process.

   You will also have purely decorative images, meant as a means of livening up an otherwise dull slide. Pick these much later, when you do your final tweaks.

4. **Decide on the placement of the different types of slides**

   - **Title slide (slide 1)**. This should contain the title of the presentation, your name, role, and institutional affiliation.

---

[1] Carter (2021).

12.2 A Slide Deck for a Short Scientific Presentation

- **Intro slides (slides 2, 3, 4)**. In this sequence, present your broader research question or topic, some background information, and the main question or goal for your project.
- **Home slides (slides 5, 6, 9)**. This is the structural "meeting point" for your presentation.

> **The home slide**
> A home slide is like a hall in a building. You enter the hall by the Intro corridor, and exit it by the Outro corridor. In the hall, there is a door to each of the experiments you are going to describe (Fig. 12.1).
>
>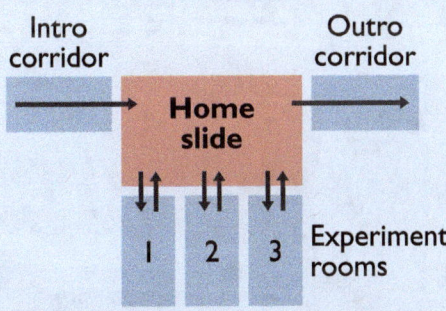
>
> **Fig. 12.1** The functionality of the home slide

- **Experiment slides (slide 7–8, and slide 10–11)**. For each experiment, present a data slide and a summary slide to give a clear breakdown of your methods.
- **Outro slides (slide 13–16)**. This end sequence includes an overall conclusion, a broad ending with contextualized results, discussion, speculation on future directions, and acknowledgements. There is also a summarizing "stay-on-the-screen" slide for the discussion afterwards (Fig. 12.2).

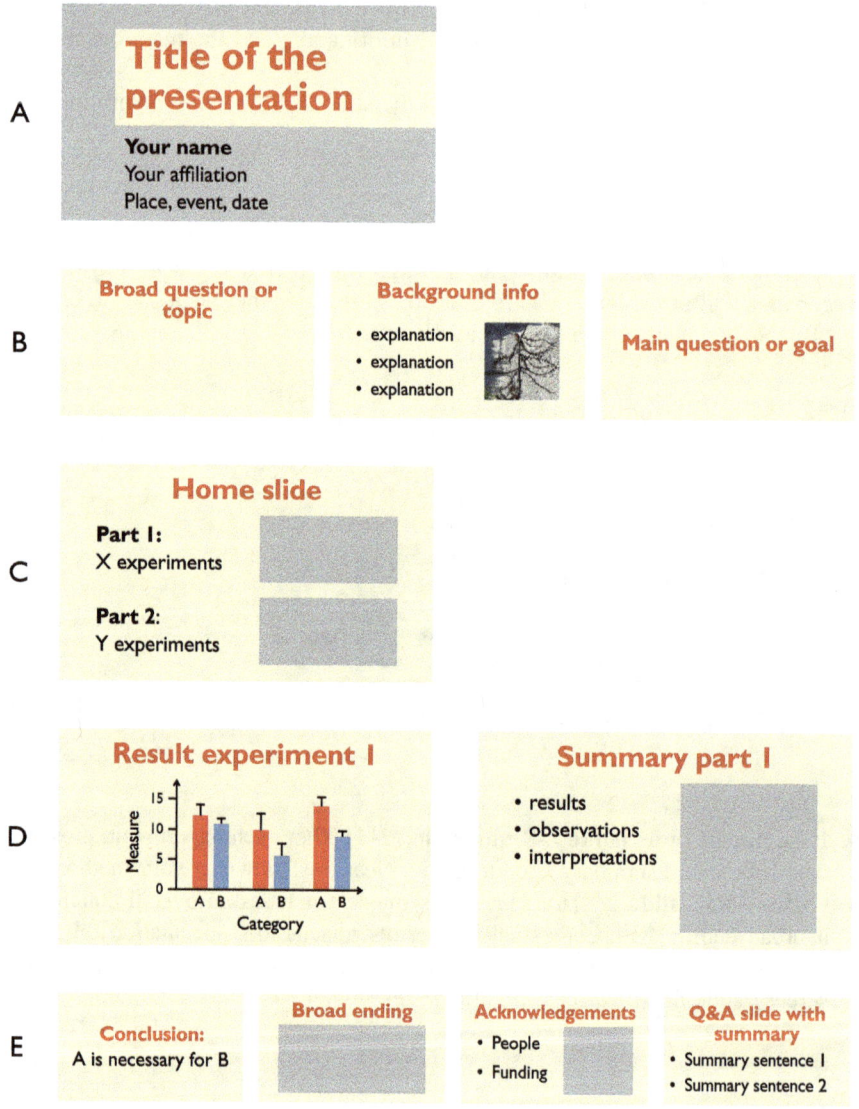

Fig. 12.2 A template for a basic, ten-minute presentation, containing (**A**) a title slide, (**B**) intro slides, (**C**) home slide, (**D**) experiment slides and (**E**) outro slides

5. **Optimize the colors and typography**

Now that we've figured out the structure of our presentation, we can think about the look and layout of each slide. Either use your university's PowerPoint template or adjust the style to your own preferences—however, we recommend that you abstain from introducing "fanciness" in the form of flamboyant themes, color gradients, unconventional typefaces, and so on. Prioritize simplicity and structure.

When it comes to text...
- **Less is more.** As soon as a slide appears in front of the audience, their natural instinct is to begin reading. The slide is thus directly competing with you for the audience's attention! To keep the focus on yourself, limit the text to a minimum. Avoid outlines and lists. Include only the few sentences that you want to emphasize, highlighting the important words of the take home message with bold letters or a bright color. Or better yet, don't add any words at all—sometimes the best option is to just have an image.
- **Choose a font style and size that's easy to see from far away.** If you're speaking in a large lecture hall or classroom, you need to make sure the people in the back can easily read the words. Sans-serif fonts, like Arial or Segoe UI, are much easier to read from afar than serif fonts, and if sized appropriately, can be read from the room's back row (see Sect. 9.6 for further guidance).

Less is also more...
- **When it comes to graphs and tables.** Researchers love graphs and tables, and put care and love into crafting them for their scientific papers. But not all are created equal, nor do they all contain the most essential information. Limit yourself to one graph per slide, so that the audience doesn't get distracted away from your speaking points. Provide an informative title to emphasize the main results illustrated by your figures. Don't use tables, unless they are really simple (say, 3×3 cells)—99% of the time the information they contain is presented more effectively in another format.
- **When it comes to photos.** Pick relevant, high resolution photos, and make them big—ideally one per slide. If you must include multiple photos in a slide, make sure they are spaced evenly, and uniform in size. When labeling a photo or figure, only label the part relevant to your talk; use a simple arrow or a bright color to highlight a specific region.
- **When it comes to the background.** Go for white (or off-white) if the presentation is projected in a bright room, and black, dark navy, or charcoal in a formal, darker auditorium, or for on-screen display. Avoid placing pictures under text—it reduces legibility.

6. **Arrange the content using the visual hierarchy**

In Sect. 9.5, we said that key information can be made highlighted with color, size, or visual weight. As you put text, figures, diagrams, and photographs into the slide, place them in an intuitive way to help the audience track the information. We also recommend grouping related elements together—but at a certain point, a slide may appear too cluttered, or too busy, in which case it is best to break up the information into multiple slides. Incorporating negative space is a high-impact way to allocate proper attention to the most critical parts of your slide.

7. **Plan how you direct attention**

   In many cases, a full picture of your results will emerge only after you have presented several graphs or figures. Displaying them all at once distracts from the particular point you are explaining through each figure.

   To direct your audience's attention, we recommend using PowerPoint or Google Slides animations[2] to reveal the figures one at a time. By the time you finish talking, all the components will be on the slide. You can also animate specific elements of a single image or graph, so that it sequentially stands out to the audience. Whether it be spotlighting a point on a line graph or zooming in on a specific structure, animations give you the ultimate control over your presentation.

Further reading

Carter, M. (2021). *Designing Science Presentations: A Visual Guide to Figures, Papers, Slides, Posters, and More.* Academic Press.
Also, take a look at the work of PowerPoint guru Echo Rivera.[3]

12.3 Designing an Effective Scientific Poster

Purpose: in general—to disseminate your science in a succinct and visual way; at conferences—to find like-minded scientists for cooperation on different levels; in the home lab—to explain projects to visitors and other contacts.
Typical situations: poster exhibitions at small and big conferences; improvised presentations in the lab corridor for visitors.
Mistakes to avoid: poor structure, messy layout and design, too much detail and lack of main messages.

To turn the poster into a truly effective communication tool, we need to clarify one thing: the poster is *not* a scientific article pinned on the wall. Instead, it is primarily a tool for networking and feedback.

---

[2] Let us be super clear: in this context, "animation" means that a figure will appear or disappear—it will NOT do several bunny hops followed by a pirouette before reaching its destination on the slide. Nobody likes over-the-top animations; they are a waste of time. Keep. Things. Simple.

[3] Rivera, E. *Create Engaging Presentations.* Echo Rivera. Retrieved August 19, 2021, from https://www.echorivera.com/.

12.3 Designing an Effective Scientific Poster

> **Your poster is NOT …**
>
> **Your poster is NOT a journal article pasted on the wall.**
> Instead, it is a billboard showcasing your ideas, projects and work in progress, sending an invitation to discuss and interact.
>
> **Your poster is NOT the complete story.**
> Instead, it is a visual abstract, a conversation starter, and a feedback generator.
>
> **Your poster is NOT the reason you're attending the conference.**
> Instead, it is a tool that complements you—as you network with, and get input from, fellow scientists.

We can illustrate this with a metaphor (Fig. 12.3). Let's say that the conference you're visiting is a lake or river, and the contacts you want to make are the fish. What's the poster then? Your fishing rod!

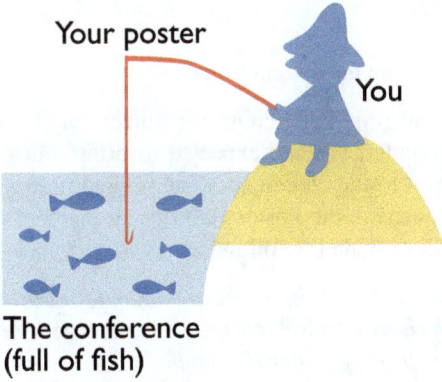

Fig. 12.3 Going fishing with a poster at a conference

Let's now go over some fundamental points to make your fishing experience a success.

1. **Set the title**

    When registering for a conference, you'll likely submit a poster title and abstract months in advance. Though you may change the content, let us be clear—the title is an extremely important part of your poster. It's often printed on the program booklet and appears on the conference website, where other attendees will peruse and preemptively decide to stop by your station—or not. You'll cite it on your CV and place it on your personal website (see Sect. 14.3 for info on how to set one up).

    A title could be written in two general ways:

- **The project-oriented title**. This kind of title describes the setup and/or the ambition of the experiment. This may be the best choice if your results are preliminary or you want to present a knowledge field, rather than the outcome of a recent experiment.

    Example: "*Effects of substance X stimulation of Y-mediated gene Z expression in ABC cell line*"

- **The conclusion-oriented title**. This kind of title summarizes your main conclusion(s).

    Example: "*Substance X downregulates protein Y-mediated gene Z expression in ABC cell line.*"

You then need to consider three things: **clarity**, **length**, and **searchability**. The title must be easy to understand and it must not take too much space on the poster. In addition, a sense of style is needed: you should know your audience, understand the context, and feel what works. Consider keywords that will make your title accessible to a broad audience.

Try writing three to five alternatives and ask colleagues for feedback to choose the best one.

2. **Pay attention to the poster instructions**

    A typical instruction from a conference organizer may look like this:

    As a poster presenter, you are expected to print your own poster (portrait; A0 size) and bring it to the conference. The printed poster should be displayed on its designated board in the Poster Hall during the course of the conference. The organizers will provide the following:

    - *a 1.0 m (width) X 2.5 m (height) board adapted for **portrait**-oriented posters in A0 format (**841 mm x 1189 mm**),*
    - *a printed card with your poster ID number,*
    - *tape, tacks and tools (scissors, staplers, etc.) for poster mounting.*

    The crucial information here is the *size* and *orientation* of the poster. Check, double-check, and triple-check this information before you start any design! Nothing is less professional than having your poster sag onto the floor because you created a portrait poster for a landscape board.

3. **Use a template**

    Let's face it—sometimes, there just isn't enough time to create a poster from scratch. It could also be that you don't feel particularly brave to break the mold and step away from the classic poster set up, but may want to try other alternatives. In these cases, templates are very useful.

    A major design choice you have to make is this:
    - Should your poster be mainly **a stand-alone presentation** (meaning it will be read while you are attending talks or mingling elsewhere)? Then aim to make it *detailed* and *complete*—like the Classic format.

"Classic" format: the tried and true

The safest bet is to take the templates used most often in poster sessions. Many universities today provide a scientific poster template that will help you swiftly design a decent poster (see Fig. 12.4). What you lose in flexibility or personal style, you gain in versatility and simplicity.

Fig. 12.4 A typical scientific poster template—this one is from Lund University,[4] southern Sweden. Note that the template makers are suggesting an inverted-pyramid-approach, starting with the conclusion in the upper left part (used with permission)

---

[4] Lund University (2020).

When you're using a pre-formatted template:

- **Start with the title, then the visuals**. The tricky part with templates is to get a good balance between visuals and text without messing up the format. Start by adding your title, your most important visuals and some placeholder text. Once you've found a layout that works, add the actual text.
- **Stick to the grid**. As soon as you start moving boxes and columns horizontally or vertically—a few millimeters here or there—the overall impression will become very messy.

- Yet if the poster serves as a **backdrop to your oral presentations** (meaning that you spend time by your poster, communicating in person with people who are passing by), then your priority is to make it *appealing* and *accessible*. For this, try the Better Poster format.

An alternative: The Better Poster format

In 2020, Mike Morrison, a PhD candidate in organizational psychology at the University of Michigan, introduced a new poster structure using his background in designing web pages.

His experience as a User Experience (UX) Designer had taught him two things. Firstly, users (any users, not just poster viewers) "skim and filter frantically, looking for the bits they need, and ignoring the rest." Secondly, users "give up the instant they feel overloaded." Based on these observations, Morrison came up with the #betterposter versions 1.0 and 2.0 (Fig. 12.5).

Morrison's idea is that the viewer approaches interesting posters and people through a three-step process.

- **Selection:** The viewer observes the posters from a distance, then identifies and selects those of interest. This can be compared to the normal behavior of web users who "tend to scan text instead of reading it closely, skipping what they perceive to be unnecessary information and hunting for what they regard as most relevant."
- **Approach**: The viewer approaches to study the selected content more closely, perhaps even reading the text.
- **Engagement:** The viewer engages and interacts with the presenter.

Note that this is similar to the behavior discussed in the O–I–A format—see Sect. 8.4.3.

12.3 Designing an Effective Scientific Poster

In the light of the "user experience" of the regular poster session attendee, Morrison opts for a template with three sections:

1. The **center banner** containing the main finding, in a large, easy-to-read typeface
2. The **need-to-know section** that contains the information found in the classic format such as the introduction, methods, and results.
3. The **nice-to-know section** contains more details or explanations for other experts in your field.

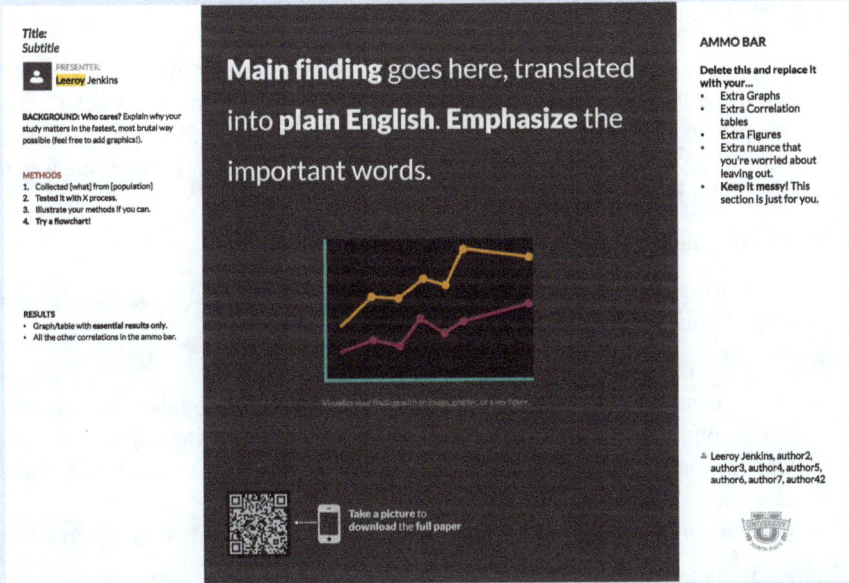

Fig. 12.5 The Better Poster layout. An adjustable PowerPoint template, along with extensive guidelines, can be found on Mike Morrison's OSF project site Better Scientific Poster[5]

4. **Select the contents**

   Regardless of the format, there are a few content modules that must exist on the poster. These modules can be shuffled around to create a story. Below are some recommendations for the different elements of the poster.

   - **Skip the abstract**. An abstract is useful on the conference website and in the conference folder, but it takes up valuable real estate on a poster. The poster in and of itself can be seen as an abstract.

---
[5] Morrison (2019).

- **Keep the aim of your study short and sweet**. Hit the ground running and do like the action stars of Hollywood: throw your viewer into the action and trust them to follow you at a high pace. An introductory line and three bullet points will probably work fine.
- **Cut the nitty-gritty details of standard methods**. Include only the information necessary to understand the results. Focus on what's unique or unusual. Better yet, think of a visual way to present the methods section (for instance, your experimental design).
- **The stars of the show are the results—choose the right ones**. These explanatory visuals, data graphs, tables and images will likely be taken from your publications. Pick the ones that communicate the major points of your findings.
- **Finish with three conclusions**. By the time you finish putting the above sections in, there's not really room for a discussion. Ask yourself "What are the three most important things that I have learned from this study?" Use the promises of the Introduction as the starting point and respond to them with three bullet points and perhaps one or two sentences about suggested follow-up studies.

5. **Refine the design**

The following are a few signposts to help guide you through design choices as you build your poster.

When it comes to layout...
- After deciding on the format, arrange your content modules so that they tell a complete story. To help your reader get to the point quicker, put the Conclusion in the upper left corner (where you start reading) and the Results in a central position.
- Let the content breathe by inserting negative space between the elements.

When it comes to text...
- THINK BIG. Because the poster will likely be printed on A0 size paper, the header font size is recommended to be > 72 point, while the body text font size > 36 pt.
- Make sure the content is legible from a distance of about 1.5 m. For best readability, we suggest a sans-serif font.
- Gradually add text—but keep it concise. Text should be left-justified.

With color, follow the KISS principle: Keep It Simple, Stupid!
- While an aesthetically pleasing poster is ideal, your main goal should be to communicate your main messages clearly. Make sure your palette isn't a distraction.
- If you have a main figure your poster centers around, try developing a color palette from the image as we describe in Sect. 9.7. Make sure it's accessible to those with visual impairments.
- When it comes to the background, prioritize legibility. High contrast is key.

Giving a poster presentation

Creating the poster was only half the battle. Now it's time to present!

Litmus test your audience

In most poster conferences you'll present at, you'll spend the majority of your time talking to people 1:1. It's a great opportunity for you to tailor your presentation. Think of your poster as a conversation starter, and take a minute to learn their background and their interest in your subject.

Play the part of a host

If you've ever been part of a small dinner party, or participated in a discussion panel, have you noticed anything about the host or the emcee? Yes, they were friendly, and perhaps they were very accommodating. Likely, they also were attentive, and answered any questions you may have had. But also—*they rarely had their backs turned to you.*

Likewise, when you present your poster, don't turn your back to your audience—not even to look at the poster. At this point of your research stage, you most likely know the ins and outs of your poster like the back of your hand.

Be friendly and open, but not pushy

Give people the space to browse through your poster, and back out if they don't find it interesting. Neither party benefits from being stuck in a forced conversation.

12.4 Producing Video Material, Such as an Abstract

Purpose: to increase visibility and raise your profile to the public and researchers in different fields.
Typical situations: social media, YouTube, journal supplements, [online] conferences.
Mistakes to avoid: lengthy videos, technical language.

Through the added benefits of social media, content consumption has never been higher than before. Inevitably, *how* information is consumed is also different. With the entrance of Tik Tok, Instagram Reels, Snapchat and YouTube, visual presentation reigns in short, abbreviated bursts.

Video abstracts fit neatly into this new style of communication. They are a quick visualization of the purpose and results that add to the understanding of a paper. The first may have been a *Cell Press* video in 2009 (now at 33 K views).[6] Most video abstracts from that era were 5–10 min long, with shots of research and specific machinery wrapped in a personalized peer-to-peer summary of an academic paper. Today, they are 3–5 min in length.

Video abstracts are extremely valuable—a 2019 study showed that among graphical abstracts, plain language summaries, published abstracts, and graphical abstracts, video abstracts scored the highest "for comprehension, understanding, enjoyment, and desired updates."[7] Comprehension was consistent across different specialties and industries, meaning that video abstracts can be effective in many scenarios. In addition, having a video is a surefire way to promote your research. Exposure and visibility can be deciding factors for receiving public or private research funding.

1. **Think like a screenwriter, not just a researcher**

   A video abstract, at its core, is a short film. No matter the budget, independent or deep-pocketed, it still requires a strong script and a storyboard.

   At its most basic, the script should answer the following questions:
   - What inspired your research?
   - How was the research conducted?
   - What results did your research reveal?

   Towards the end, there should be an invitation for the viewer to read your paper.

   Storyboarding is extremely valuable at this stage to ensure your visual guide fits with the script. This will also help in determining the visuals and locations you want to show in the video.

   Obtain all the necessary permissions. If you want to film in a lab, you should communicate this to the lab's PI in advance. Importantly, mention how long the filming will take, so the other researchers in the lab can plan their experiments ahead if needed. If the filming involves any patients or participants, make sure you obtain written permissions from them to use the footage.

2. **Keep your tone light and friendly**

   Avoid technical jargon at all costs. While this may be difficult, the purpose of video abstracts is to increase accessibility and interest in your research. It won't just be other researchers viewing the video, but also students, journalists, or industry partners. Plus, if people want to know more, they can read the full details in the actual paper.

---

[6] Enard et al. (2009).
[7] Bredbenner and Simon (2019).

3. **Develop a great hook**

   The first 8 s are crucial to grab your audience's attention. Consider these methods:

   - **The Curiosity Gap.** The curiosity gap—the link between what an average viewer knows and doesn't know—triggers the viewer's curiosity (for instance, *"Flamingos come in a range of bright pinks and reds—but did you know why?"*). This initial interest can already be piqued via the title and thumbnail of the video.

     The essence of this hook is based on the assumption that the viewer has some sort of knowledge of your research topic. By building on the familiarity, you can delay the video's main idea.

     Examples of great informative video creators that use this hook are Vox's *Earworm*[8] series and Bloomberg Quicktakes' *Hello World with Ashlee Vance*[9] series.

     Warning: A pitfall with this method is that it is easy to lose or confuse the viewer by making them feel they aren't intelligent. Then, any chance of engagement goes out the window.

   - **Aristotle's Technique.** This classic 3-step method gets straight to the point of your research:

     1. **Tell the audience what they will learn**, that is, the biggest highlights of your research.
     2. **Show the audience what they were told they will learn**. Talk about the background information and the methods used. Of course, do not use complicated language.
     3. **Recap** the key takeaways.

     Aristotle's technique is a staple in the beauty, tech and IT sectors since it is so effective at communicating value.

     If your research is heavy in technicalities, or the method is the focal point, use this hook.

   - **The S.I.C. Storytelling Formula.** Let's face it—everyone likes a good story—even in a video abstract.

     S.I.C. stands for **Situation, Interest,** and **Conclusion**. It is a storytelling formula that helps develop an attractive narrative around your research.

     1. First, explain the **situation:** the background and the "players" of your research that are connected to the problem you are trying to answer.
     2. Then, draw **interest:** flesh out the problem and how the "players" dealt with it, and how the situation developed.

---

[8] Vox (2019).
[9] Bloomberg Originals (2019).

3. Finally, **conclude** with how the problem was answered, and what "lesson" was learned through the players.

Great examples of this hook can be seen in TED-Ed's animation videos, VICE's *Story Of* series,[10] or Wendover Productions' videos on how things work.[11]

- **Click-bait**. Similar to the curiosity gap, a click bait shows a preview that promises the viewer something interesting or shocking. The more dramatic it is, the more interesting viewers believe the video will be. Examples of click-bait videos are far and wide on YouTube and other platforms—in fact, it is probably what causes you to watch the videos you do.

    Warning: A click-bait is a promise, and so if not satisfied, the viewer will immediately lose interest.

    Start with the most shocking or most relevant results of your research, and some background info to hook the viewer.

    Then, begin your actual story. Click-baiting can be easily merged with the S. I. C. Storytelling Formula. After the preview, you can further interest your viewer in the narrative of your research.

- **Humor**. The largest benefit of communicating science via video is that your enthusiasm for the topic can be shared. People can get a peek into a world that they may never experience first-hand.

    With this in mind, humor is an excellent tool to introduce your research in an approachable way. Here are some ways to use humor in the hook:

    I. **Situational irony**

    Situational irony is defined as actions or events that have the opposite result of what is expected or intended.

    To use situational irony in a video abstract, take what people expect from your research topic—and flip it!

    For example, Hank Green, American science communicator, YouTuber and internet personality often uses situational irony in his TikTok videos to break his viewers' expectations.

    II. **Self-Deprecation**

    The perception that science is difficult or only for "smart" people can make communicating science a challenge from the get-go. There's also the idea that whoever is communicating the science is probably an unapproachable authority. In order to break this wall, you need to show your viewer that you, too, are human.

---

[10] VICE (2005).
[11] Wendover Productions (2010).

Using self-deprecating humor shows that you are willing to laugh at yourself, which will lower the layperson's guard. Excellent examples of this are shown by the hosts of the popular *SciShow* YouTube videos.[12]

4. **Keep track of the time**

A typical video abstract is 4 min long. For social media, aim for 2 min and 20 s. To keep your thoughts in order, writing a loose script is recommended. Keep in mind that a minute of speaking time translates to 120 to 150 words, so limit your script to 750 words or less.

5. **Try to outsource filming whenever you can**

First, see if your institution has a Press Office Department or a Public Relations Officer. This is likely the same department that helps you write a press release (see Sect. 11.5). They may lend filming and recording equipment, edit your footage, and even recommend content creators to collaborate with. They may already have experience creating video abstracts, and if so, will have tips and tricks to help you prepare. Trust that the department knows how to help you communicate your science.

If no such resource is available, do not fret! Creating a video abstract is still possible; all it takes is some cleverness and creativity.

6. **Establish your budget**

The quality of a video abstract may differ depending on the budget, but it's not to say that a fantastic one requires a lot of money. After all, video abstracts will be shared on social media, where often the more polished a video looks, the more unapproachable the creator seems. Creating an excellent piece is possible at all budget levels—high, medium or low (Table 12.1).

Table 12.1 The options available for productions with a range of budgets

Budget type	Description
High	With a large budget, you may have access to state-of-the-art equipment, or perhaps the ability to outsource the production. The sky's the limit to your creativity and execution. Elements that can be considered in such a video include: • author(s) introduced and talking on camera, aided by a teleprompter, • high quality graphics, whether 3D or 2D, • complicated animations, • re-enactment of laboratory methods, • any related footage in connection to the research

(continued)

---

[12] SciShow (2011).

(continued)

Budget type	Description
Medium to low	If the budget does not allow for professional cameras, it is possible to have a mostly animated video abstract with a clear and understandable voice over. In addition, check to see if it is possible to rent cameras, lighting and audio equipment, etc. with a local AV rental company. It is definitely cheaper than buying expensive cameras. Elements that can be considered in such a video are: • author(s) introduced and talking on camera, • clever use of **high quality** animated graphics, • any related footage in connection to the research
Low	Regardless of funding or previous filmmaking experience, a great video abstract can be made with the tools available to you. The level of technology in this day and age means that state-of-the-art equipment isn't a must. In fact, some current smartphones are able to record in 4 K or high-quality videos. However, using a smartphone does limit the use of specialized graphics, animations, or other video enhancements, meaning that keeping and capturing the viewer's attention is important. For this, there is also downloadable free software such as Windows Media Player or iMovie, and for making designs, there are options like Inkscape. Solutions to technical aspects such as sound should be considered. Another solution is Fiverr, a freelance service site, where it is possible to hire someone to professionally produce your audio/video. Elements that can be considered in such a video are: • author(s) introduced and talking on camera, • overlaying images, • a focus on the narrative of the story

Conveying science with animation

Three tips from Tor-Martin Austad, Visual Lab, Oslo, Norway. Tor-Martin Austad is an animator and biologist, specialized in explaining science to the public by using visuals and animation.

- Base the visual style and composition on your message, who the audience are, and what type of engagement you are trying to invoke.
- Animation is tedious work. Make it easier for yourself by compressing the content to short but informative sentences. Don't try to animate every word you want to convey. Instead, invest in animated scenes that say a lot and can be further built upon.
- Decide how the information is being conveyed in addition to the animation. Through text, infographics or a speaking presentation/voiceover? The animation and text/audio should work together as supplementary tools to grasp the concept, not as disturbing elements creating cognitive conflict and confusion.

Acknowledgements Thank you to Tor-Martin Austad and Mike Morrison for contributing to this chapter.

References

Bloomberg Originals. (2019). *Bloomberg*. YouTube. Retrieved August 19, 2021, from https://www.youtube.com/@business

Bredbenner, K., & Simon, S. M. (2019). Video abstracts and plain language summaries are more effective than graphical abstracts and published abstracts. *PLoS ONE, 14*(11), e0224697. https://doi.org/10.1371/journal.pone.0224697

Carter, M. (2021). *Designing science presentations: A visual guide to figures, papers, slides, posters, and more*. Academic Press.

Enard, W., Gehre, S., Hammerschmidt, K., Hölter, S. M., Blass, T., Somel, M., Brückner, M. K., Schreiweis, C., Winter, C., Sohr, R., Becker, L., Wiebe, V., Nickel, B., Giger, T., Müller, U., Groszer, M., Adler, T., Aguilar, A., Bolle, I., & Calzada-Wack, J. (2009). A humanized version of Foxp2 affects cortico-basal ganglia circuits in mice. *Cell, 137*(5), 961–971. https://doi.org/10.1016/j.cell.2009.03.041

Lund University. (2020). *Download templates and communication tools*. Lund University, Staff Pages. Retrieved August 15, 2020, from https://www.staff.lu.se/support-and-tools/communication-and-graphic-profile/download-templates-and-communication-tools

Morrison, M. (2013, August 19). *Mike Morrison, PhD*. YouTube. Retrieved July 20, 2021, from http://www.youtube.com/@MikeMorrisonPhD

Morrison, M. (2019). *Better scientific poster*. Osf.io. Retrieved August 25, 2020, from https://osf.io/ef53g/

Rivera, E. *Create engaging presentations*. Echo Rivera. Retrieved August 19, 2021, from https://www.echorivera.com/

SciShow. (2011, October 21). *SciShow—YouTube*. YouTube. Retrieved August 19, 2021, from https://www.youtube.com/@SciShow

VICE. (2005). *VICE—YouTube*. YouTube. Retrieved August 19, 2021, from https://www.youtube.com/@VICE

Vox. (2019). *Vox*. YouTube. Retrieved August 19, 2021, from https://www.youtube.com/@Vox

Wendover Productions. (2010). *Wendover productions—YouTube*. YouTube. Retrieved August 19, 2021, from https://www.youtube.com/@Wendoverproductions

Two-Way Communication in Various Contexts

13.1 Introduction

Imagine being the person who can "talk to countrymen by the manners of countrymen, and with the scholarly person in Latin." This skill was praised in a 1905 classical poem by Swedish writer Erik Axel Karlfeldt[1]—and it is particularly valid for specialists whose expertise, while essential, can often create a barrier between them and people less informed about their subject.

But even within expert communities, cultural divides arise based on background, profession, priorities, and values. This was famously raised by British physical chemist and fiction writer C. P. Snow. In May 1959, he addressed the Senate House at Cambridge University in a lecture that soon became a book: *The Two Cultures and the Scientific Revolution.*[2] Snow argued that intellectual life had split into two camps: science and the humanities.

Although Snow's book is still worth studying, we note that today's scientists often experience a broader range of "Two Cultures":

- academic science versus industry R&D
- academic science versus startup businesses
- science versus media
- science versus educational system
- science versus policy making
- science versus activism
- science versus religion
- … and so on

---

[1] "… *han talar med bönder på böndernas sätt/men med lärde män på latin.*" From the poem *Sång efter skördeanden* by Erik Axel Karlfeldt (1864–1931; posthumously awarded the Nobel Prize in 1931).

[2] Snow (1962)

What's important for the communicating scientist is that the formulas that work in the safe context of science departments may fail totally in the dwellings of The Other Cultures. So, when stepping into the entrance of an industry building, a startup village, a high school, or a national agency, remember the words Dorothy says to her dog Toto in the movie *The Wizard of Oz* (1939): *"Toto, I've a feeling we're not in Kansas anymore."*

13.2 Networking

> **Purpose**: to make the most of the few precious hours you have at various events for knowledge exchange, networking and gaining inspiration.
> **Typical situations**: scientific conferences, networking events.
> **Mistakes to avoid**: staying in your social comfort zone; forgetting to have a plan; chatting about your own stuff without listening to the other party.

Having a strong network of contacts is absolutely priceless—they serve as your hivemind, gateway for career opportunities, or even social support. Attending conferences and other specialist events is the perfect occasion for building and strengthening that network, especially since most have designated networking sessions (either explicitly scheduled as such, or disguised as coffee breaks, pre-conference drinks, and so on). While many people find the idea of forcibly chatting to strangers highly uncomfortable, the right mindset, preparation and attitude make it much less awkward and much more useful. Here we show how to make the most of such events.

1. **Ask yourself: "Why am I going?"**

   As with all communication activities, you should be clear about the purpose. Here are some potential examples:

   - Connect/interact with a certain professional for a certain purpose.
   - Build a network of partners, decision makers, influencers and gatekeepers.
   - Open up new areas (intellectual, business, geographical…) and engage in new activities.
   - Consolidate your wider professional network.
   - Promote a discovery, an idea, an invention or a product.
   - Make a name of yourself as an expert or influencer in your field.
   - Find new friends and people you can trust.

2. **Research the context and the people you will meet**

   What do you know about the event, venue, schedule and participants? Think about different aspects, for instance:

- What is the infrastructure and logistics? Is it a designated networking event, or a coffee break where everyone is mostly focused on their caffeine intake and glucose levels? How crowded is it likely to be?
- What is the timeframe? The big fish (keynote speakers, important politicians, celebrities) will usually only make a short appearance, so if your main goal for this event is to meet and interact specifically with them, act fast.
- How formal or informal is the event? This will dictate what you might want to wear, and how you would approach people you don't know.
- Is the social structure more hierarchical or flat? Who do you know already? How can they help you reach the person you have come here to meet? Resist the urge to only hang out with your friends—you can do that any time; networking is more focused and purposeful.
- Are there any aspects of the language or cultural context that you should be aware of[3]?
- Are there any tribes or ideologies that you need to know about? Will talking to one group or clique get in the way of meeting anyone?

3. **Go through your own strengths and weaknesses**

    Do a little SWOT (Strengths-Weaknesses-Opportunities-Threats) analysis on yourself—use the template from Table 13.1. What sort of a person are you—an introvert or extrovert? Spontaneous or a perfectionist? Ambitious or laid back? Target-oriented or open to serendipity? Do you have people skills? Language skills? Humor? What qualities could help you at this event?

Table 13.1 A template of a personal SWOT analysis, which can be useful to help you identify your goals and needs. Fill each square of the table with a list of items relevant to each component; however many you need to reach a thorough conclusion

Strengths	*Weaknesses*
What are your skills, natural strengths, relevant bits of knowledge and expertise?	What are the weaknesses you could improve on to achieve your goals? Identify any skills gaps, relevant training that you are missing, or perhaps any personality traits that could get in the way.
Opportunities	*Threats*
What external opportunities are available to you? Are new areas opening in your field? Is there a new funding call for your specialism? Is your area becoming a bit of a hot topic across the sector?	What external factors might put you at a disadvantage? Is your field very competitive? Are you worried about precarious contracts, or disruption within the sector?

---

[3] It took Joanna longer than she'd like to admit to figure out that when Brits say "Let's do dinner sometime!", they really mean "see you NEVER."

Come up with some conversation starters (this is particularly useful if you are somewhat shy and find it difficult to strike a conversation) and practice them—on a friend, or even just in front of the mirror.

Plan your persona, your entry, your strategy, and your exit.

4. **Prepare an elevator pitch**

   We have included a recipe for this in Sect. 10.2. Remember the fundamental questions:

   - What do you do?
   - What problem do you solve?
   - How are you different?
   - Why should I care?

5. **Know things**

   After all, this is why people will want to talk to you. Bear in mind, you don't need to be a top expert in the field (though if you are, well done!)—you can bring an interesting angle to the conversation by leaning on your background experiences. For instance, you could be a decent particle physicist, but your real strength lies in public engagement skills, and knowing how to obtain funding for citizen science projects based around physics.

6. **Give more than you take**

   Giving and receiving is at the core of networking. Center your interactions on the question "How can I help you?"; listen to what the other party needs, and see if you can find a way to assist. This attitude is necessary for professional success—positioning yourself as the person who knows things and is helpful will draw more people towards you. Being generous and humble pays off.

7. **Be clear about the next step**

   You have had a nice chat, both parties seem to have gotten something out of it; now is the time to do a quick assessment of the future of the formed relationship. Is it a contact for a particular purpose (new country, subject, research topic...)? Part of a wider network? Starting point for cooperation? How will you take it forward—will you discuss the next steps via email, or meet up face to face? Jot this down in a notebook or on your phone, so it doesn't escape your memory the moment you leave the venue.

8. **Always do a follow-up**

   Many junior people who mustered up the courage to chat up a senior keynote or important professor miss the opportunity to "close the deal" by forgetting to follow up. If someone has given you their contact details, use them! Send a friendly greeting via email, or add them to your LinkedIn network. Confirm whatever you agreed upon, send links to information, schedule a meeting. Be proactive—this is the moment to tighten the relationship and make it more personal and worthwhile.

A Mingling Protocol
Arrive well-prepared

- Set goals: "Who do I hope to meet, and why?"
- Analyze the context.
- Dress appropriately—either blend in or stand out[4] (but make it a conscious choice).
- Manage your blood sugar.
- Bring contact cards and a notebook.
- Practice some conversation starters.

Observe

- Look for people you know (and ask them to introduce you to their network).
- Look for the people you want to talk to—and figure out who could introduce you to them (or just use one of those conversation starters you prepared).
- Look for people who seem a bit lonely.
- Consider the timeframe and prioritize strategically.

Engage

- Explain who you are, what you do, and what you're looking for.
- Listen actively with the ambition to connect, serve, and help.
- Point the person to valuable information.
- Take notes.

Disengage

- Quit while you're ahead (don't overstay your welcome in the conversation). Let your politeness guide you.
- Exchange contact information.
- Summarize your interaction and the next steps.

---

[4] If you want, have a "whatsit"—an interesting brooch, glasses or t-shirt, which gives others a pretext to strike a conversation with you.

Follow up

- Connect (email, LinkedIn).
- Confirm anything you agreed upon, schedule any planned interactions.

13.3 Presenting in an Industry Setting

Purpose: to initiate cooperation on different levels—e.g. data sharing, methods and instrumentation, clinical studies, prototypes, et cetera.
Typical situations: startup business communication activities; networking.
Mistakes to avoid: forgetting the priorities of the industry; not explaining the practical value or economic potential; getting lost in details.

Let's begin with a story—travel with us to a distant galaxy, where Planet Academia and Planet Industry coexist.

Most of the time, the inhabitants of Planet Academia are engrossed in their own pursuits. Occasionally, they must journey across the galaxy to meet the inhabitants of Planet Industry. These industrious people have a Universal Translation Gadget at their disposal, but unfortunately, the Cultural Understanding Unit has not been updated in years. As a result, communication between these two groups is not always straightforward.

So, while the natives of Planet Academia are eager to present all the details, the inhabitants of Planet Industry impatiently demand the conclusions. And while the Academians prefer to focus on pure theories and ideas, the Industrians are determined to discuss the grease-stained realities, introducing topics like production issues, markets, money, and profitability—that seem "vulgar" to anyone from Planet Academia. Nevertheless, for the prosperity and security of the galaxy, these two groups need to achieve smooth cooperation through clear communication.

What is the lesson from this little allegory? Understanding is possible, as long as the interacting professionals from the two cultures try to—sincerely and generously—understand each other's mindsets and priorities. The main points are summarized in Table 13.2.

13.3 Presenting in an Industry Setting

Table 13.2 Priorities of academic and industry audiences. Based on Mahak, Francine: presenting your research to an industry audience

Academic audiences want to know:	Industry audiences want to know:
• How rigorous is your research methodology? • How is your presentation expanding what I know or contradicting what I believe? • Are your findings interesting? Reproducible? *Academic institutions must show intellectual merit, win grant funding for research, etc*	• What are your findings? • How do they apply to what we do? • What is reasonable to expect, and by when? • How will your skills and experience contribute to our success? • What are the risks and uncertainties? *Industry organizations must stay ahead of their competition to remain viable—and not waste time on projects that take too long to implement or are too risky*[5]

1. **Make sure you understand what's expected from you**

   Whatever you do, don't grab that slide deck you normally use to present to your science peers. Remember: you're now a space traveler, visiting Planet Industry. Instead, start the process by having an upfront conversation with the person who invited you. Use the questions below as starting points, and take notes while you listen.[6]

   - What is the purpose of the event? What is the preferred outcome? What would success look like for me as a presenter?
   - Who will attend? How much do they know about my subject?
   - What areas of my research are they interested in?
   - Are there any specific conclusions/experiments/figures they want me to highlight?
   - What are the company's objectives? What are their differentiators in the market, and what are they or their competitors doing better?
   - Who are the key interested parties and beneficiaries?

   Remember: it is far from certain that your main focus aligns exactly with their interest. To serve your audience, you may have to go back in time to earlier projects, or talk about something you have given a lower priority. Also, if your goal is commercialization, be clear about what you want the next steps to be.

2. **Try to get a glimpse of the people and the culture of the workplace**

   Start by looking at the company's website. Then, for a more critical view, dive into discussion groups and social media. The next step is to be curious about the staff you will meet, and their background. Let this be the second part of the conversation you're having with the person who invited you.

---

[5] Mahak (2019, September 4)
[6] That's just general advice: you should always take notes when doing research for your planning and concept creation.

- Do they come from the same team, or from a range of departments?
- How many of them have a science background? Are there any former colleagues from your own research field or university?
- Will they be generally positive about what I present, or will there be criticism from parts of the audience?

3. **Plan and prepare a front-loaded presentation**

   When preparing your talk, we strongly recommend that you change the way you tell your story. Industry audiences will generally assume that you've followed the scientific process meticulously. Because of that, the normal IMRaD format is not very interesting to them, nor is an experiment-based approach.

   Instead, they are curious about how you can make yourself and science useful—especially inside their area of activities. This is why you should focus on the following:

   - **Highlight the results first**: What useful—perhaps even profitable!—knowledge did your research uncover?
   - **Explain their value**: How do your findings relate directly to the activities of the company and the departments represented in the room?
   - **Show practical applications**: What solutions, methods, and practical lab skills did you come up with to obtain these results?
   - **Provide key takeaways**: What are the actionable messages?

   Start your talk with the findings, then explain why these are important and how you got them, and circle back to the results and next steps.

4. **Be open about unsolved problems and roadblocks**

   R&D in industry is more about problem-solving, adaptation, and optimization than it is about totally new concepts and innovation from scratch. As a matter of fact, it is in the lifeblood of engineers to focus on problems and impediments and how to get rid of them.[7]

   Therefore, idealizing your research process and making it seem smoother than it is will only slow down the communication. Issues great and small must and will be addressed sooner or later—definitely sooner if the engineers actually find your concept interesting.

5. **Be prepared to serve information you didn't have before**

   If you have invented something that carries commercial potential, you may suddenly find yourself venturing into uncharted territory—areas like marketing, production and sustainability. Questions may look like this:

---

[7] Joanna *loves* working with engineers—unlike scientists, who are people with problems, engineers are people with solutions.

- What does the competition look like?
- How feasible is it to put the product into large-scale production?
- How do you solve waste, sustainability, or security issues?
- How far is it from being market-ready? What kind of testing has been done?
- How big is the specific problem you have solved? For example, if you have found quicker electric vehicle chargers, look at the EV market impeded by slow charging—not climate change in general.

6. **Be prepared to improvise**

   You may find that industry culture is not as polite and patient as academia. It's not only about being outspoken and to the point. It may be possible that they reschedule with short notice or suddenly give you twenty minutes instead of the hour you expected. Just swallow your frustration and see it as a challenge to solve any issues in the smoothest way; be proud of your professional compliance!

7. **Respect their rules**

   High-tech companies are very concerned about security—both regarding their intellectual property and the health and welfare of employees and guests. Be a smart visitor: listen to instructions, follow the footsteps of the people guiding you, don't go exploring on your own, don't read stuff from whiteboards or printouts, and—in general—keep away from things that seem unfamiliar.

13.4 Public Engagement

> **Purpose**: to inform and inspire the public; to inform a wider audience about your findings; to engage non-specialists in your research.
> **Typical situations**: anything from public talks and science festivals to citizen science events.
> **Mistakes to avoid**: underestimating or being dismissive of your audience, not matching the type of activity or the demographic to your aims, forgetting to evaluate.

Science communication, outreach, public engagement, science dissemination… Do these concepts mean the same thing? While they sound similar—and many science people do indeed use them interchangeably—the differences between them will become increasingly apparent as we immerse ourselves in the broad field of "academics communicating with a *general* (non-expert) audience."

Science communication, often abbreviated to **SciComm**, is predominantly a one-way process, focusing on the dissemination of knowledge—or the "transmission and translation of scientific results and their impacts"[8]—from experts to non-experts. It may encompass anything from scientists providing commentary on current developments, writing "explainer" pieces, holding open lectures or making social media memes that clarify scientific concepts.

Similarly, **outreach** is a one-way process, with the caveat that it is typically aimed at schools. For instance, science outreach may be done to increase interest in science subjects among children, or boost the number of university applications to STEMM programs.

On the other hand, **public engagement** (sometimes abbreviated to **PER**—Public Engagement with Research) is based on a dialogue *with* the audience rather than a monologue directed *at* them. The National Co-ordinating Centre for Public Engagement—a brilliant British hub of PER-related wisdom—defines it as follows:

"Public engagement describes the myriad of ways in which the activity and benefits of higher education and research can be shared with the public. Engagement is by definition a two-way process, involving interaction and listening, with the goal of generating mutual benefit."[9]

As we can see, public engagement, though a bit of an umbrella concept, has two key aspects: it is (a) interactive and participatory, and (b) mutually beneficial to both parties involved (Table 13.3).

Table 13.3 There are a few concepts used to describe situations when science is communicated to non-scientists

Term	Direction of process	Focus	Relationship
Science communication, SciComm	One way: expert to non-expert	Dissemination of knowledge	Producer of knowledge → consumer of knowledge
Outreach	One way: expert to non-expert	Inspiring and educating young people	Producer of knowledge → consumer of knowledge
Public engagement, PER	Two way: expert and non-expert in dialogue	Share outcome of science with the public	Both parties are co-producers of ideas or knowledge

In the text, we have done our best to find definitions that represent how different types of activities are currently categorized. This table serves as a summary

Both science communication and public engagement have gained significant momentum since the 1990s. In many countries, they are already embedded into the

---

[8] Nerghes et al. (2022).
[9] National Co-ordinating Centre for Public Engagement (2020).

research strategies of funding bodies and universities. And there certainly are good reasons for academics to interact with "normal people." Some key ones include:

- **Accountability**. Most research is paid for by public funding. It is only fair that the public knows and understands how their money is spent.
- **Trust**. Public understanding of the scientific process, the nature of enquiry, and scientists as people, builds a stronger relationship with researchers. In today's world, where everyone has access to a sea of information—of varying quality—it is important that expertise is trusted and valued. In the long run, such trust helps to reduce anti-science views, and promotes a dialogue instead of a combative approach.
- **Commitment to public good**. Much like increasing cultural capital, increasing scientific capital is a good thing. While not every child needs to be a career scientist, it is beneficial for everyone to have a deeper understanding of how science works, an analytical approach to new information, and confidence in gaining new knowledge.
- **Grant requirements**. Many funding bodies now require project results to be disseminated into the wider world beyond academia. This is your chance to shine! All too often panicked academics put down a desperate, last-minute dissemination attempt along the lines of "I'll give a talk about this project at my kids' school"—but, unless your target audience consists of children, this is not a well-focused or thought-out engagement activity.
- **Impact**. Demonstrating that your research has a far-reaching impact comes in handy during appraisals—both on personal and institutional levels.
- **Informing your project**. Citizen science, where non-experts contribute to data collection or analysis—for instance through recording species' sightings or examining camera trap photos—can help scale up, or speed up, scientific efforts. The public can also be consulted about the direction in which research is going; this is particularly common in biomedical work, where patients suffering from a particular disease are often actively involved in the development of research projects.
- **…because it's fun**? Yep, this is a valid reason as well. In fact, for many of us, it's what started the whole thing.

1. **Establish your goals**

    The choice of engagement type will depend on your goals—so, as with other communication endeavors, take some time to think through your activity and how it aligns with your objectives. A great starting point is to consider a few key questions before choosing (and committing to) a particular type of engagement activity:

    - **What do I want out of this?** Why am I spending time and energy on this activity? Will it benefit my research project? My career prospects? My personal desires?

- **What do I want from the recipients of my message?** Am I looking for new directions for my work, for more data, for funding? Do I want to raise awareness, educate, inspire? Am I hoping to recruit study participants or students for my program? Or, in short, do I want to disseminate knowledge, consult with the public, or collaborate with them on a project?

2. **Establish who your audience is**

   There is no such thing as a "general" audience; define your demographic—and be as specific as possible. Picture the exact person you are addressing—is it a teenager in Boston, or a mother of a premature baby in Nairobi? What is their age, gender, background and geographical location? What do they find engaging and relevant?

   Are you communicating with someone who has an existing interest in your research area (e.g., patients suffering from a disease that you study; students on an astronomy summer school visiting your lab; amateur gardeners cultivating pollinator-friendly plants…), or are you reaching out to new, most likely uninterested people? What is their attention span like—will they be able to spare two minutes or half an hour? Where are they based—at a designated science festival, or in a supermarket? To reach a broader audience, consider exploring unexpected venues—for instance public toilets (where the audience can be engaged for 2 min), or airports (where people wait around for a long time).

   Pro tip: if, for whatever reason, you have no idea what audience to expect at an event, a good rule of thumb is to pitch at the level of an intelligent teenager with a mild interest in the subject.

3. **Choose an activity best suited to your goals**

   Not every type of engagement is suitable for all purposes. You have just answered the first two big questions about **what** you want and from **whom**, now it is time to think about **how** to achieve it.

   > **Matching the PER Activity to Your Objectives**
   > If your aims are:
   >
   > - **To inform and inspire the public…**
   >   - Consider: participation in science festivals; interactive talks and shows; digital engagement, for instance via films and animations.
   >   - Examples: shows; open talks and events (Science Slams, Pint of Science, Soapbox Science); science communication competitions (Three Minute Thesis, FameLab, or, if you're brave, Dance Your Ph.D.); digital and in-person engagement with schools (Lecturers Without Borders; I'm a Scientist, Get me out of here!).
   > - **To consult and listen to public views**, including concerns or new perspectives…

13.4 Public Engagement

- Consider: public debates; online consultations; panels and user-groups.
- Examples: Patient and Public Involvement groups (especially in the medical sciences); working with organizations that already have access to your target demographic.
- **To collaborate with the public** on data collection or analysis, developing future research directions or defining policy outcomes…
  - Consider: Crowd-sourcing; citizen science apps and platforms; co-production of knowledge.
  - Examples: platforms such as Zooniverse or iNaturalist; data collecting events like BioBlitz or the Big Garden BirdWatch.

4. **Figure out the constraints**

Every engagement activity works within a framework of constraints—try to pinpoint them in advance. Some events, such as science festivals, may ask you to provide a risk assessment of your activity; if you plan on working with children, you may need to undergo a criminal history check beforehand. Aside from the legal paperwork, there are also the more mundane aspects—such as the amount of space you have at the venue, or the noise levels. What resources might you need? Will you have enough for the expected footfall?

Also, think about your strengths. Do you like talking to people, or would you prefer writing or making memes? Do you love working with kids, or have a soft spot for senior audiences (who, by the way, make wonderful engagement groups—they have plenty of time, and are more relaxed and more open-minded than a lot of younger people)? While it's good to get out of your comfort zone, don't force yourself to do an activity that makes you feel very uncomfortable.

5. **Build up the right attitude**

Once you have everything planned out, think about your audience once again. Ask yourself:

- **What's in it for them?** Why should they spend their time listening to you? Are you entertaining, informative, useful, relevant…? Remember, they are doing you a favor by giving up a bit of their time to listen to what you have to say. Make it worth their while.
- **What's the key takeaway?** Clarify what is the one key message your audiences should remember from the activity. Consider having it written down in a form of a giveaway; even a simple one such as a flier, card, or sticker could do the job.

Most importantly, never be dismissive. You may be the expert, but for all you know there could be a retired professor of your field in the audience. What is more, people from outside a discipline have the most amazing and unexpected insights thanks to a fresh take on the subject.

6. **Run the activity**

   Public engagement can be a tough job. A lot of the time, if the audience is truly "general," you are a mix of a babysitter, preacher and used car salesman. It takes a certain type of personality. Having said that, it can be incredibly rewarding—so enjoy every minute (and of course, as discussed in Sect. 7.6.2, expect the unexpected)!

7. **Evaluate**

   You've spent some time disseminating your science, or engaging the public—but how can you know that they got your message? Think about how you can measure the impact of your activity; this is essential for determining your effectiveness and considering what needs to be changed in the future.

   There are plenty of evaluation resources available, but it can be as basic as having boxes, marked "I learned something new," "I already knew this," "I found this exciting," into which people throw beads. You can obtain feedback on post-its—if they are shaped like something relevant (such as leaves that you can put on a poster of a tree for a botany-related project), all the better. To see what knowledge stuck, do a mini-quiz—with little prizes, like candy or stickers. For online engagement, check the analytics and the metrics.

   You could also start the project with a formative evaluation, which acts a bit like a pilot study to assess the initial idea and troubleshoot it.

   Use the evaluation results to build your future projects, and evidence your success.

> **Starting a Career in SciComm**
>
> So, doing science communication and public engagement has been so much fun that you are now considering taking your relationship with it to the next level. How should you proceed if you want to make the bold switch from a lab environment to a career in science communication?
>
> Our first tip may not be what you want to hear: **don't quit your dreary postdoc engagement—not yet!** Firstly, it is exceptionally tough to find a SciComm position without documented experience and a portfolio. Secondly, if you choose to be a freelancer, it may also be tough to find enough clients for decent cash flow. And thirdly, the science community can be unforgiving to those who try some other career move and then come back knocking; leaving the experimental research world often means burning bridges.
>
> The good news, however, is this: in a networked and globalized world, it's very easy to get started on the side. Start learning, start connecting, start

producing—and when opportunity knocks and the time is right, you'll be ready.

In the meantime, here is a bit of guidance to get you on your way.

Engage in your university's outreach activities. Seek out the teams responsible for public engagement, outreach, or even communication and marketing, and have a chat with them. There may be regular events that require volunteers and participants; offer to write a piece for the departmental website; volunteer to take part in a promotional video.

Engage in whatever is happening in your neck of the woods. Does your university have a science center, or do they cooperate with a museum? Then they're probably on the lookout for volunteer guides. Are you part of a specialist group or a learned society? Help them run an event at a fair or school.

Sign up for global events and competitions. See what is available to you and have a go—what's the worst that can happen?

Take courses in whatever feels useful. There are many online and in-person courses (many probably even offered by your institution) focused on science communication and public engagement. Remember, though, that other courses may be just as useful for your career: journalism, podcast production, illustration, and so on.

Start producing and broadcasting content. Be it your own blog, podcast, video series or social media channel—do it your own way, find your favorite themes, and build your network while learning and practicing, practicing and learning. The bolder you are online, the sooner the world understands that you're worth listening to.

Find ways to get published. Approach your local newspaper or a national magazine with an idea for an article that addresses something their audience is interested in. The closer to your expert area, the easier it is to convince any editor that you're the right person to cover the specific subject. Try your hand at writing for *The Conversation*—it's a great starting point for learning how to work with an editor (plus the editors there are used to working with academics).

Further science communication and public engagement resources. You might find some of the following communities and organizations useful:

- Public engagement and science communication bodies, such as the National Co-ordinating Centre for Public Engagement,[10] British Interactive Group,[11] Australian Science Communicators[12] or British Science Association[13]
- Professional bodies (for instance, Royal Society; American Society for Cell Biology; British Ecological Society, and so on)
- Universities—many have general resources and specific tips on how to run public engagement projects
- Local support: there may be a designated public engagement officer at your institution—find them and get in touch.

13.5 Building a Relationship with Mainstream Media

Purpose: to take charge of the discourse; to create a symbiotic relationship with journalists you can trust.
Typical situations: you want to make your voice heard in the public sphere; you would like to speak to or write for the media more directly.
Mistakes to avoid: not giving the journalist clear, useful and interesting information; being unreliable, unavailable or unresponsive.

Working with anyone (but the media in particular) is based on trust and mutual benefits. In this section, we go through how to commence, develop, maintain and make the most of a relationship with journalists and editors.[14]

---

[10] National Co-ordinating Centre for Public Engagement (2020).
[11] British Interactive Group (2020).
[12] Australian Science Communicators. (2024, March 6).
[13] British Science Association (2020).
[14] A journalist's job is to gather information, conduct interviews or investigations, and create content. An editor is responsible for content selection, development and organization—or the overall production of the publication. In some cases the roles may overlap, but the rule of thumb is: if you want someone to produce material about your research, contact a journalist; if you want to create something yourself, contact an editor.

1. **Have an idea**[15]

    This includes the idea of what to write or say, as well as an idea of the form in which you'd like to present it. Media outlets offer a vast choice of formats—opinion pieces, expert comments, explainers, and more. Ask yourself: What do you feel confident and comfortable doing? Would you rather write or speak?

    Once you have made up your mind, decide what sorts of media outlets would be most suitable for you, and make sure your story is interesting and relevant to them. Choose your target carefully—a shotgun strategy, including the world and his dog, is *not* recommended! To know your way around diverse media options, be a consumer, not just a producer. Read, watch, listen to the outlets that you might want to contribute to—this will allow you to fit into the style of the publication, understand the form and what they are after.

2. **Make an inventory of contacts**

    In marketing, salesmen talk about *warm* calls or *cold* calls. The first category contains people who know you, or at least feel acquainted with you. The second category is made up of strangers. Obviously, it is much harder to get through to the second category: you have to make them understand who you are, what you're after and whether they can trust you. You should therefore aim to turn your emails, texts and phone calls into warm calls—or at least semi-warm.

    When you go through your address book or your LinkedIn catalog, look for contacts in this order:

    - Media people you know.
    - Friends who can help you get in touch with journalists.
    - Media people you don't know yet.

    Once you have worked with someone, make a note of how the interaction went, and whether it is worth repeating.

3. **Write a media pitch**

    If you are making a general announcement about something newsworthy, and hope to reach as many journalists as possible, write a press release (see Sect. 11.5). If, however, you want to target a specific outlet or reporter, and get them to report on your news, or if you want to submit your own piece to a particular editor, you are better off sending a media pitch. Pitches tend to be tailored, and therefore are likely to get greater traction.

    While a press release is a good way to send out a longer statement that expressly communicates your news, a media pitch explains why a certain angle or story is newsworthy and appropriate for a specific outlet or audience.

---

[15] The Austrian journalist and satirist Karl Kraus famously—and unkindly—said that journalists have "no ideas and the ability to express them."

> **Sample Pitching Email**
>
> Keep the introductory email short—not more than 200 words. Try to approach editors directly, especially if their email is available on the outlet's website. Also, get introduced by colleagues who have written there.
>
> Below is a sample of what you might include in your pitching message.
>
> **Subject**
>
> *Submitting a guest essay on* [TOPIC] ← for a cold call.
> or
> [FRIEND'S NAME] *sent me your way* ← for a warm call.
>
> **Body**
>
> *Dear* [EDITOR],
>
> *I'm a* [POSITION] *at* [INSTITUTION] *who specializes in* [FIELD], *and would like to make an argument about* [TOPIC]. *I work on* [ASPECT OF TOPIC] *and have previously written for* [MEDIA].
>
> If the piece you propose is short, include it in its entirety; you need to show what you are capable of to build the relationship and instill confidence. Give a brief description of the argument in the body of the email before attaching your work.
>
> For longer pieces, add
>
> - what you will write,
> - why should they publish it (how is it relevant to their audience).
>
> **Timeframe**
>
> If the piece is time sensitive, put a timeframe on it:
> *If I don't hear from you within* [24 or 48 h] *I will assume it's a no.*
>
> Otherwise check in once a week, then after a while give a clear timeframe. Being politely persistent is a good approach.
>
> Unlike press releases, never submit an opinion or letter to multiple places at once. It will be very awkward if all accept it.

As you write, imagine that the addressee is someone NOT looking to say "yes." In fact, the model response is no response. A "no" is still kind of ok, especially if the rejection includes anything good or useful—after all, the editor looked at it and has taken the time to respond. A "yes" is, obviously, a success.

Remember that there is a lot of discretion involved in the decision process, and good pieces may be passed because of oversaturation, especially in larger media outlets. Don't get too discouraged (or too encouraged!) by initial results.

4. **Build on an existing relationship**

   Once you have a positive experience working with a journalist or editor, don't squander the trust you have built. Be responsive, receptive, and of course,

13.5 Building a Relationship with Mainstream Media

deliver interesting content—doing so will gain you the label of "reliable contact," and make you the person journalists and editors default to when they need an expert comment or opinion. In fact, if you get to know them well, and they trust you to be a person who delivers the goods (speaks in an engaging way, writes well, and so on), they will say yes to most of your ideas.[16]

- **Take the initiative**. For interviews and materials prepared as evergreens (not linked to a particular event), and especially if the journalist knows your background only vaguely, you could suggest concrete questions or angles. Usually they are more than happy for you to take the lead. Journalists vary, however—some like to do solid research and ask concrete questions, others prefer to let their guest do the talking more freely.
- **Travel**. Many broadcasters will be based in a major city—often the capital. If you are from out of town, but traveling to where the journalists are located, give them a heads-up. They might have a free slot to record a segment; it provides new content and saves them the journey.
- **Adapt to their work dynamics**, **not yours**. They may disappear for two weeks and then suddenly show up and give you a couple of hours to seek out information or fact-check something. This is how the media works, so just roll with it.
- **Rise above political or cultural divides**. Build a relationship with journalists even if they work for outlets you don't read or agree with, such as a particularly vicious tabloid, or a TV station promoting a political agenda you oppose. Don't alienate that segment of the population. Accepting a left-field invitation and expanding your viewership can build you a reputation of an apolitical, objective expert, especially when the topic you discuss is not politicized, but simply focused on science. Besides, if you don't do it, someone else will—and they might peddle an agenda that's not objective at all.

5. **Capitalize on the trust**

One of the biggest fears stopping scientists from being more present in mainstream media is that their words will be twisted and they will embarrass themselves in front of their peers (and the rest of the world... but mainly their peers). To alleviate that fear, ask your media contact if you may take a look at the article before publishing.[17] A polite request to fact-check should

---

[16] This is definitely true, in our experience, for smaller media outlets, or those in a language that's not English (and the pool of guests is, naturally, smaller).

[17] In some countries, such as Poland, a journalist is legally obliged to present (if asked) a draft for "authorization" before publishing—that is, to get the interviewee to confirm that the words on paper were, in fact, theirs. Most journalists hate this requirement, since it adds another step to the content production process, and also is an opportunity for people to cut out the juiciest bits of the interview. Still, it acts as a bit of a security blanket for the scientists who worry that their statements might be misconstrued. In other countries, however, the opposite is true—showing the final text to the source can be perceived as violating the newspaper's code of conduct.

be fine, provided you get back to the journalist very quickly (usually within a few hours), as they operate on a tight deadline. While getting the facts straight lies in the interest of both parties, you are the person who risks their reputation if a mistake is made. After all, when seeing a dodgy or wrong statement in an interview, the readers will think "idiot scientist," not "sloppy writer."

Still, most of the time the best approach is to serve a good story and subject expertise, and then leave the journalists to do their job their way. If your approach to offering suggestions and ideas starts feeling too intrusive, you may provoke them to drop the whole initiative.

6. **What if things go wrong?**

Make the most out of your positive relationship with the journalist and get the facts straight ahead of the publication—there is not much you can do once the text is out (and incorrect). It is important to remember that you will not always be entitled to final editorial control over what is written or shown, so try not to sweat the small stuff if the general gist is accurate.

What can you do if you don't like the final text or program?

- Not much, unless there is a strictly factual error—in which case a correction can be issued (though, in all honesty, very few people will read that).
- Live with it. Joanna once had a photoshoot for a magazine, to accompany an interview with her. The photographer was great, and most of the pictures turned out really well—except, in her opinion, the photo the editor picked for the cover. You can't do a lot about a case like this; it's not factually wrong, so there's no way (or real reason) to correct it. You just have to swallow your vanity and look away when you see an ugly picture of yourself on a magazine stand.
- If the experience wasn't great, avoid working with that particular journalist or editor again—but don't let it put you off media in general.

13.6 Addressing Skeptical Audiences

Purpose: to convince, rather than inform, your addressees (activism, if you want to call it that).
Typical situations: open lectures, public engagement, social media presence.
Mistakes to avoid: being disrespectful, relying on facts alone, not listening.

Depending on your research topic, you may end up interacting with people who are not neutral towards your work, but, in fact, hold a strong bias against you—often

13.6 Addressing Skeptical Audiences

before you even open your mouth. This holds especially true if you are researching a topic considered controversial (or "controversial" to the general public, not amongst scientists). Such topics include:

- vaccine development
- climate change
- using animals in research
- genetically modified organisms
- cats killing wildlife
- trophy hunting.

Here, we provide some advice on how to handle communication with skeptical audiences.

1. **Remind yourself of how humans think and interact**

    Before you go any further, do a recap of Chap. 5 on the psychology of communication, to review confirmation bias or cognitive dissonance (*"I love my cat and I love birds, and knowing that roaming cats kill birds is causing me mental discomfort"*).

    Also, be aware of how society, culture, and technology shapes our opinions and the way we express them. First of all, because information is widely available, everyone can be a self-proclaimed expert on the back of a bit of googling. Secondly, societies have become increasingly polarized—in no small part due to social media—and we increasingly find ourselves surrounded by bubbles of like-minded people, taking a stance against those who hold different views. Online interactions leave little room for nuance; furthermore, there is constant pressure to have an opinion: not taking a stance is, in itself, a stance.

2. **Be realistic about your goals**

    Take a moment to reflect on your goal. Is it to *inform*, or to *convince*?

    As science people, we generally feel more comfortable informing: we limit ourselves to talking or writing about interesting bits of research, hoping that our audiences make the most of it. There will be times, however, when—for a range of reasons—informing may not be enough. It could be because you would like to change your audience's behavior, change a policy, change a public attitude—or get support for your cause in a different way.

    Do remember, for your own peace of mind, that you will not convince everyone—and that's okay. Focus on those open to learning and move on from those who are not.

3. **Get to know your audience and their motivations**

    Find out as much as you can about your audience. Who are they? What's their background? What experiences shaped them? What is the attitude they currently present? Where are their preconceptions or prejudice coming from? Try to unpick the nature of the biases you come up against.

Don't label. Someone who protests against vehicle restrictions in low traffic neighborhoods might not necessarily be a "motorhead"—instead, they could be a non-driver frustrated by the restriction's negative impact on public transport. Or perhaps someone is not inherently opposed to an idea, but there are practicalities that get in the way? For instance, a colleague of Joanna's who was researching vaccine hesitancy in the Orthodox Jewish communities in the UK realized that the "hesitant" behavior (that is, not vaccinating kids) doesn't always stem from beliefs or principles, but from the fact that vaccination sites were frequently not accessible for mums of multiple kids, wielding double pushchairs.

4. **Find common ground; build trust**

The most useful and important point is to **listen**. Hear out the other party's reasons, worries and doubts, and acknowledge these feelings. Often, frustrations with scientists and experts stem from a perception that they are dismissive or arrogant. Demonstrate humility and focus on shared goals.

Remember that ultimately, we probably have more in common than we think: we all want safe, happy lives for ourselves and our families; we just have different visions of how to achieve that goal.

5. **What doesn't work**

There are a number of things that, perhaps counterintuitively, don't work when it comes to changing a stance or behavior.

- Relying solely on facts. Surprisingly for some, pure facts are not enough! Remember that informing is not the same as changing opinion. And repeating facts at an increasing volume definitely doesn't work.
- Lecturing, talking down at people, being patronizing. It makes your audience feel small and stupid—not a good starting point for building trust.
- Yelling and mocking. This is alienating, not helpful. Memes and scorn might be an outlet for personal frustration; they are amusing to the people "on your team"—but they won't convince anyone to change opinions. Imagine the reverse situation: would you ever say "Oh yes, these derisive comments aimed at my fundamental beliefs *totally* made me abandon them!"?
- Saying "I told you so"—even if you are right.
- Nagging. Repeating without meaningful dialogue risks you sounding like a broken record. People will start to avoid you, or, at best, label you as the crazy evangelist for your cause.

6. **What to try instead**

- Be polite and listen to arguments. Treat everyone as if you would like to be treated.
- Acknowledge emotions.
- Build relationships and social capital. Bring in human stories, not dry facts.

- Leave space and time for people to change their minds—it's not an easy process, and changing a core belief can be painful.
- Always let the opponent save face and keep their dignity. If they decided to alter their beliefs, don't rub it in their faces; they made a big step and it should be acknowledged.
- If you feel competent, you could employ Socratic reasoning. Ask thoughtful questions that encourage reflection.

> **Reflect on Your Own Experience**
> Think back to the last time you changed your opinion, belief or behavior.
> - **What** did it take?
> - **Who** did it take? Who do you trust?
> - **How long** did it take?
>
> and…
> When did you last change somebody's mind? On what? How?

7. Look after your own wellbeing

It is easy to advise patience, listening, and building trust. But we are all humans—and we all get tired and exasperated sometimes. The feeling that you are constantly fighting your corner is exhausting; pushback or even accusations ("all scientists are corrupt and research only serves big pharma," to name just one such barb) can be very demotivating. Don't let that get you down; take a break from such interactions if you need to, try not to get too overwhelmed. If you vent, do so in private—trust is easy to break and difficult to rebuild.

Finally, know when to say "enough," and put energy into something else. Sometimes you need to ask yourself: do you want to be right, or do you want to be friends?

Acknowledgements Thank you to George Economides, Joseph Fridman, and Natalia Osica for contributing to this chapter.

References

Australian Science Communicators. (2024, March 6). *Australian science communicators.* Australian Science Communicators. Retrieved October 8, 2020, from https://www.asc.asn.au/

British Interactive Group. *BIG STEM Communicators Network - Home.* BIG: STEM Communicators Network. Retrieved October 8, 2020, from https://www.big.uk.com/

British Science Association. *British science association.* British Science Association. Retrieved October 8, 2020, from https://www.britishscienceassociation.org/

Mahak, F. (2019, September 4). *Presenting your research to an industry audience.* Inside Higher Ed. Retrieved October 9, 2021, from https://www.insidehighered.com/advice/2019/09/05/how-present-effectively-industry-audience-opinion

National Co-ordinating Centre for Public Engagement. *National Co-ordinating centre for public engagement.* NCCPE. Retrieved October 8, 2020, from https://www.publicengagement.ac.uk/

National Co-ordinating Centre for Public Engagement. *Introducing public engagement.* National Co-Ordinating Centre for Public Engagement. Retrieved October 8, 2020, from https://www.publicengagement.ac.uk/introducing-public-engagement

Nerghes, A., Mulder, B., & Lee, J.-S. (2022). Dissemination or participation? Exploring scientists' definitions and science communication goals in the Netherlands. *PLoS ONE, 17*(12), e0277677. https://doi.org/10.1371/journal.pone.0277677

Snow, C. P. (1962). *The two cultures and the scientific revolution.* Cambridge University Press.

14 Setting up and Executing a Plan for Your Online Presence

14.1 Introduction

Having a strong online presence is generally beneficial for your career—but maintaining it can take up a lot of time and energy. So is it worth the hassle?

While it is not mandatory to curate an online persona, there are several advantages to doing so.

1. **Control**. Due to the nature of the modern world, you are, in all likelihood, already present online (through your work, events you attend, friends you have)—whether you want to be or not. You might as well do so on your own terms.

2. **Opportunities**. Many opportunities come via social media—for work, public engagement or collaboration. The three authors of this book are speaking from experience, since they have all been approached by journalists, employers and future colleagues via social media.[1] It is also a way of engaging with celebrities or influencers who might have strong opinions on topics they are not experts on—but which you happen to specialize in.

3. **Support**. A broad network can be incredibly helpful—for support or as a hivemind. Want recommendations for a field notebook app? Trying to access bird collections from eighteenth century Amazon? Ask on social media, in the ecology or museum circles.

4. **Centralization**. As you change jobs and affiliations, you will also change email addresses and official website profiles. It is therefore useful to build your own

---

[1] As a matter of fact, this book project started that way.

website—or store contacts in a more centralized place, to keep track of them even if your work circumstances change. The flip side, of course, is being reliant on the whims of the social media provider.

5. **Exposure**. If our—academic—currency is research article citations, then an online presence certainly pays off. Advertising papers on social media does increase their citation numbers.[2] After all, scientists are people, too, and they get their information from all kinds of sources.

If you are, indeed, considering maintaining a more structured online presence, ask yourself the same question you would before starting any communication activity: What are your objectives? What do you want to achieve? Defining your goals will help you focus and refine your online presence—and manage your audience's expectations.

Are you aiming to...

- Highlight your research?
- Create a space where you can be easily found?
- Find a job?
- Network?
- Meet journalists?
- Do public engagement?
- Become an influencer?
- Simply enjoy hanging out on social media (perfectly fine, too)?

Or perhaps there's something else—or a combination of factors.

Finally, determine what sort of a commitment you are willing to make. Are you prepared to maintain a regular presence on social media, or would you prefer to build a website that might only get a refresh once a year? Or perhaps both?

The path of least resistance would be to put together a LinkedIn profile, treating it as your online CV. In some countries (such as Sweden, Olle says), a LinkedIn profile is an absolute necessity for job hunting. While this may not be the case in other parts of the world, we believe that having a comprehensive outline of your professional life online can be very beneficial. A LinkedIn profile places your current work in a broader context, allows you to curate and highlight your achievements, and can facilitate future introductions. It's a one-time setup, with updates needed only when you change jobs or start new projects.

In a similar fashion, you can create a personal website. This requires a bit of time and skill, but once set up, tweaking the contents should not be too onerous. We will cover this process in Sect. 14.3.

---

[2] Some references to back this claim, from fields as diverse as (a) coloproctology (Jeong 2019), (b) communication (Özkent 2022) and (c) ecology (Peoples 2016). In fact, (d) social media attention can be used as a predictor of a study's impact (Sathianathen 2020).

However, if you have the headspace, stamina and motivation for a more interactive experience—whether for networking, exchanging ideas, influencing others, or sharing your research—social media is the best way to go.

14.2 Using Social Media Effectively

> **Purpose**: to disseminate your work to a wider audience; public engagement; networking.
> **Typical situations**: using any online platforms.
> **Mistakes to avoid**: being unfocused and unpredictable, being self-centered, being boring.

A helpful approach to using social media effectively is based on a key marketing principle—the Golden Triangle. The idea is simple: for a marketing strategy to be effective, it must be based on three components—Market, Message, and Media. The market is your audience, the message is what you want them to remember, and the media is the platform or channel through which you will reach them. Identifying the latter component is based on the two former ones—plus, we will add a fourth factor: Means, or the level of commitment you are willing to sign up for.

In the prehistoric days—i.e., late-2000s—when social media were just emerging, choice of medium centered on the three Ps: use LinkedIn for Professional, Twitter for Political and Facebook for Personal activities. Much has changed since then, including the birth of many new platforms and the altered character and usability of the old ones. LinkedIn is still used for professional activity (though less for academia and more for businesses and start-ups). Twitter—or X, or what's left of it—can be a political tool, but is equally great for science communication, creating parallel discussion streams at academic conferences, and it also makes a decent source of discipline-specific news—if you curate your timeline well. Alternatives like Threads or Bluesky play a similar role. Facebook has lost much of its appeal in the Western world, but is still very strong globally. The functionality lines blur—for example, Joanna has received job offers via Facebook and Twitter, and started sci comm projects via LinkedIn. And then of course there are Instagram, TikTok, Mastodon, Snapchat and all the others!

Once you've set the goals for your online presence, use the Golden Triangle (plus the one extra component) to determine the optimal channel for your activity.

1. **What's your *Market*?**

    Who is your audience? And no, don't answer "everyone." Just because "everyone" is on social media (actually, they aren't), it does not mean that they will instantaneously become your followers. In particular, ask yourself:

- **What age are they?** Older people are more likely to use Facebook, younger—TikTok and Snapchat. This will naturally change in a few years, as new channels emerge.
- **Where are they based?** In many countries, Facebook is by far the most dominant social medium, if not the most popular website/app altogether. Equally, there are countries that ban particular social media platforms. Are you targeting a specific area or language group? Read up on what is on the rise in the region.
- **What do they find interesting?** Does your audience have a genuine reason to check you out (as is the case with employers/recruiters, academics in your field, or real-life friends), or is it a purely transactional relationship (content producer to content consumer)? Through this question, we're trying to establish whether the followers will come to you, or whether you need to actively attract them and keep them engaged, and how much effort this will require.

2. **What's your *Message*?**

What type of content do you intend to produce? This question relates to both the format and the substance of your posts.

- **Picture, text, video, audio?** Are you a keen photographer, and does your research lend itself to something very visual? Instagram is great. Are you more of a writer? Facebook, Reddit or Medium might be the way to go. Comfortable with videos? Try TikTok or YouTube. You can also create content in a more independent set up (e.g., blog posts on your personal website) and link to it on all your social outlets.
- **What are your key messages?** When your audience follows you, they need to know what to expect. If you were to distill your content focus and vibe into one sentence, what would it be? For example: "I am well-educated, have relevant experience, am looking for work," "I am a researcher in field A and I study topic B," "I will tell you why XYZ is the most interesting thing on earth." Of course, you may have the occasional stray post going off-topic, but it is important to manage your audience's expectations, and keep your focus clear.
- **Why should anyone follow you?** If your goal is to engage a wider, non-obvious audience, you need to attract them with your humor, energy, or style. As an exercise, think of the least relevant and interesting thing for you—what would make you follow it on social media? Would it be because of a personal relationship (supporting a friend), because the account is entertaining, because it came recommended by someone you trust—or maybe there was another reason? Joanna never had much love for geology, but thanks to a friend whose passion for the subject is nothing short of amazing, she has slowly warmed up to it.

- And, at the end of the day, **what do you want your audience to do?** What is the direct goal of your activity? Is there a call to action, a project they should contribute to, knowledge that they should gain…? The world is listening—what do you want it to hear?

3. **What are your *Means*?**

   The final factor is based on the resources at your disposal, in particular how much time you are willing to spend refining your online presence. It is good to make a steady commitment at the start—especially if you are building an audience from scratch. Aim to post regularly—every day, every week—and don't limit yourself to only displaying your own content. Be a community member.

> **How to be a "good citizen" of the social media community**
> To build a stronger, more meaningful social media community around you, try to engage beyond the transactional producer–consumer relationship.
>
> - Don't limit yourself to posting content and expecting likes and follows. Engage, comment, react, repost, discuss, encourage. Do unto others as you would have them do unto you.
> - Use a tit-for-tat strategy—build relationships by engaging with other users' content.
> - Don't waste other people's time just because you can.
> - Tag people, show appreciation for their work.

Building and maintaining a community takes time—not always a lot, but with regular social media presence comes regular commitment. Before making that commitment, consider establishing goalposts: what will determine your success? Periodically evaluate whether honing your social media persona is, in fact, worth your time. Set yourself a timeframe—say, half a year—as well as a goal (such as a certain number of followers, post statistics and so on—use the analytics function of your profile), and reflect, as honestly as possible, on your progress.

> **CASE STUDY: Institutional social media**
> If you are one of the younger members of a research group, you may be asked by your PI to set up a departmental/group/institute social media account. Before you say yes (and we appreciate that this may be an offer you can't refuse), consider the time commitment involved.
>
> Ask yourself—and the PI:
>
> - Why do we need this account? Is it because "everyone has one nowadays"? If so, resist at all costs. If, however, the purpose is strategic, like

public engagement or broadening institutional impact, then it is worth considering.
- Who is your audience? As we know, "everyone" is not a good answer. Think about it this way: if a similar institute from elsewhere was on social media, would you follow it?
- Who will generate the content? Will you be on your own, or will others in the group contribute? What is the daily commitment of each of these people—and if they are attending to social media, who will do their day-to-day academic jobs? Is there enough going on in the group to make content creation easy? There is nothing worse than a sad, abandoned and unloved social media account (if you have one of those, just disable it temporarily—at least until better days come).
- How often would content be posted? To grow your audience, you need a steady commitment—posting at least once every couple of days, plus interacting, commenting, sharing, etc. Will you have time for this?
- Are there any existing tools that could do the job for you? For example, if you have a Nature paper coming out in a couple of months, it might be best to inform your departmental/university communications officer and ask to post about it from an institutional account, rather than frantically trying to create a social media account from scratch. An established university account will also have more followers than you could muster at a few months' notice. Remember: there's no point building an entire restaurant just because you want dinner.

Interestingly, an institutional social media account is very useful for *internal* communications, since the users with a vested interest in following such an account are the ones affiliated with the institution itself.

Once you have identified your purpose, audience, content, and resources, you are good to go. Regardless of the choices you went with, here are a few general tips that might come in handy when you are in doubt over whether to post something or not.

- Think: why am I posting this?
- Stay focused.
- Stay polite and kind.
- Listen, broadcast, respond.
- Don't vent, rant, offend (yes, future employers do check applicants' social media presence).
- Posts with media (photos, gifs, videos) get more traction. Use alt text descriptions (photos) and captions (video) for accessibility.
- Make use of topical days, hashtags (#inktober), events (#MuseumWeek), or trends.

14.2 Using Social Media Effectively

- Avoid humblebragging—an absolute plague on LinkedIn. Starting posts with "I am humbled to have been awarded...," or "It is my great honor and pleasure to announce..." is boring and self-absorbed. Instead, try leading with a question, or a statement relevant to your audience.
- If in doubt, consult your institutional communications officer (if there is one).
- Create your social media policy.

> **Sample social media policy**
> A social media policy is particularly useful if you are sharing account responsibilities within a team; it ensures everyone is on the same page.
>
> - Only post content you will be happy with two days later.
> - Focus on key messages, not politics (unless, of course, your core messages link to politics).
> - Avoid posting anything inflammatory.
> - Don't respond to trolls.
> - Act as if your boss is watching.
> - Don't post pictures of anyone who hasn't consented to it—and yes, this includes your own children.
>
> What would your social media policy look like?

Disclaimer

In this chapter, we have described the benefits of using social media, but there are, of course, caveats as well. The main ones center on privacy and security. Social media companies may track your browsing history to try to profile you, and your personal data may be sold to third parties—advertisers, political campaigners, etc.—or be used to train artificial intelligence. On top of that, engagement is often driven by outrage-based algorithms, leading to a more polarized society.

You have been warned!

14.3 Creating a Personal Website

> **Purpose**: to create a curated online space showcasing your experience, achievements, and interests.
> **Typical situations**: establishing an online presence.
> **Mistakes to avoid**: lacking a clear purpose, using poor-quality images, neglecting user experience.

In the introduction to this chapter we highlighted the benefits of curating your web presence. Here, we take a look at how to create an online space that showcases your work and interests in an optimal way.

1. **Consider what is already available to you**

   Before we go any further, let's take a quick look at **your profile on the institutional website**. This is probably the most important outlet for your online presence, for two reasons: (1) academic institutions tend to be positioned at the top of web searches, and (2) institutional affiliation lends credibility. Therefore, your profile is the first thing people find when they search for your name—so make a good impression.

   - Keep your page updated. Some organizations allow you to edit the content yourself, others do it through an administrator. Find out who is in charge of the updates and make friends with them.
   - Ensure your profile picture is professional and high-resolution (in particular, avoid images cropped from a group photo, especially if parts of other people are still visible).

   While a profile on a departmental website is a great thing to have, it will only exist as long as you are affiliated with that department. The current lifestyle of a science professional often involves short-term contracts, frequent institutional changes, multiple affiliations, and possibly being involved in "side hustles" such as spinout companies or science communication. It might not be appropriate to advertise your tangential activities on an institutional website, especially if they are not directly relevant to your main job—or present competing interests.

   Conversely, while a LinkedIn profile is extremely useful, its layout does not accommodate non-linear or multiple parallel careers (there is also no space for featuring a portfolio, should you need it). You might also have a GitHub repository, a blog, a YouTube channel, a podcast, or showcase your photography on Instagram or Flickr. Therefore, we highly recommend creating an online space where you feature all your experiences, achievements, and interests.

14.3 Creating a Personal Website

2. **Establish your needs and goals**

   When developing your personal website, start by defining your goals. Consider how you want to portray yourself, as this will dictate the site's layout, tone of voice, types of photos and other content.

   - **Content scope**: Will your site be strictly academic, or will it include other work? If the latter, will you feature this work on the website itself, or link to an external site (for instance, a spinout company)?
   - **Functionality**: What features will your site need—a gallery, comment box, blog, online shop?
   - **Update frequency**: How often are you likely to update the site? If not often, avoid sections like "news" or "blog"; leaving them untouched for a while will make your poor website look outdated.

3. **Choose the technical components**

   To create a website, you will need three things[3]:

   A domain. This is the address that you see in your browser's search bar, for instance www.crastina.se.[4] Keep it simple, short and relevant; avoid numbers to prevent ambiguity ("Is it *one* or *1*?"). Domains can be purchased from Domain.com or Gandi.

   A web server. This is where your website is hosted and made accessible to the world.

   A website. This is the structure, layout and contents of the actual site—either built from scratch (in HTML, CSS and JavaScript), or via a web builder (editing software).

   You can go about the process in multiple ways, but be aware that when it comes to creating a website, you can only pick two out of the following: (a) cheap, (b) easy, and (c) professional-looking. What do we mean by this?

   - For a **cheap and easy solution**, use a free web builder that provides hosting (for instance, Wix, Weebly or GoDaddy have a free plan). It's a good starting point, but without payment, you will not have a fully customized domain name—plus there will likely be ads on the sidebars of your page.
   - For an **easy and professional-looking** option, select a premium web builder plan (such as Wix, Hostinger or Squarespace), with multiple templates, a drag-and-drop editor, and widgets for a range of functionalities. The professional appearance and user-friendly editing environment are probably worth the cost.
   - For a **cheap and professional-looking** choice, you can have a go at building the website from scratch (for instance with a content management system

---

[3] Ryan and Willett (2019).
[4] Crastina (2016).

such as Wordpress), and find a host. It will give you creative control but has a steeper learning curve than the other two scenarios.

4. **Plan the layout**

    Whatever the shape, your website will need:

    (a) A homepage: the landing page that shares key information about you, and invites the audience to explore.
    (b) An about page: this provides the details about you as a person.
    (c) A contact page: for users to get in touch.

    You can add other pages, such as a blog or a portfolio. Based on your goals, you might modify the "About" page to an "Academic CV" page, and add a section for "Public Engagement" or "Consultancy." Get inspiration from colleagues with similar career paths and see how they structure their sites. You could even ask what software they used.

    Sit down with a piece of paper and draw or list the components you need. Include the must-haves and the nice-to-haves, and prioritize your work accordingly. Optimize the user experience by minimizing the number of clicks needed to access content. Avoid nesting menus within menus, and keep sections compact to prevent excessive scrolling.

5. **Gather the assets**

    Collect photos, videos, text files and other materials. Ensure all images are of high quality—chances are they will be viewed on a large screen. To reduce loading times, encode images as JPG or WebP. Have a selection of portrait, landscape, and square photos; some will work better on phones, and others on laptops—make sure the photo that looks great on your computer doesn't end up with a cut-off head when viewed on a tablet.

    Think about where you will be placing text, how much of it you want, and what tone it should be written in; then write the copy.

6. **Get building**

    Have a look at the templates available to you—select the one that best suits your needs. Then tweak it to personalize the site.

    Preview your website on multiple devices (laptop, tablet, phone), and via multiple browsers (especially Firefox). The viewing experience will differ based on how the site is accessed—and while you will probably be designing the site on a computer, most visitors will likely view it on their phones.

    Use alt text for image accessibility: provide a short description of each image (this can usually be done with GenAI), which will allow people who use screen readers to have a more complete experience. For the same reason, avoid generic button labels, such as "click here," opting instead for "learn more about my research" or anything else that provides context.[5]

Ask a few friends to review your site. Are there any broken links[6]? Is everything loading properly? Implement changes and publish.

7. **Tell the world**

Wait, I've just built a website to promote my work and now I need to promote that very website? We're afraid so! Put it on your social media; link to it from your institutional site (if nobody objects); shout about it from the rooftops. Add it to the bottom of your conference bio. Engrave it on your dog's collar tag, if you think it might help!

To see if your marketing efforts paid off, use analytics tools.[7] These can help you track traffic, identify which pages and keywords are getting most hits, and where your audience is coming from.

Acknowledgements Thank you to Ian Preston for contributing to this chapter.

References

Brokenlinkcheck.com. (2010). *Free broken link checker—Online tool*. Online broken link checker. Retrieved November 10, 2020, from https://www.brokenlinkcheck.com/

Crastina. (2016). *Crastina—Science communication for early career scientists—A networking platform*. Crastina. Retrieved November 10, 2020, from http://www.crastina.se

Jeong, J. W., Kim, M. J., Oh, H.-K., Jeong, S., Kim, M. H., Cho, J. R., Kim, D. W., & Kang, S. B. (2019). The impact of social media on citation rates in coloproctology. *Colorectal Disease, 21*(10), 1175–1182. https://doi.org/10.1111/codi.14719

Özkent, Y. (2022). Social media usage to share information in communication journals: An analysis of social media activity and article citations. *PLoS ONE, 17*(2). https://doi.org/10.1371/journal.pone.0263725

Peoples, B. K., Midway, S. R., Sackett, D., Lynch, A., & Cooney, P. B. (2016). Twitter predicts citation rates of ecological research. *PLoS ONE, 11*(11), e0166570. https://doi.org/10.1371/journal.pone.0166570

Plausible. (2020). *Plausible Analytics*. Plausible Analytics. Retrieved November 10, 2020, from https://plausible.io/

Ryan, E., & Willett, N. (2019, November). *How to build a website: The step-by-step guide to easy setup*. Website Builder Expert. Retrieved November 10, 2020, from https://www.websitebuilderexpert.com/building-websites/

Sathianathen, N. J., Lane R., III, Murphy, D. G., Loeb, S., Bakker, C., Lamb, A. D., & Weight, C. J. (2020). Social media coverage of scientific articles immediately after publication predicts subsequent citations—#SoME_Impact score: Observational analysis. *Journal of Medical Internet Research, 22*(4), e12288. https://doi.org/10.2196/12288

---

[5] You can do an accessibility audit of your site using tools such as (WebAIM 2021).

[6] Use an automated broken link finder like (Brokenlinkcheck.com 2010).

[7] For privacy-respecting analytics, use plausible.io.

WebAIM. (2021). *WAVE web accessibility tool*. WAVE. Retrieved November 10, 2020, from https://wave.webaim.org/

Epilogue: A Communication Manifesto for Science

Throughout this book, we have done our best to provide insightful analysis, sensible overviews, useful procedures, and handy tricks of the trade. Our journey involved a lot of reading, researching, and discussions, leading to many new discoveries. More importantly, we have also tightened up sloppy preconceptions, structured half-baked thoughts, and cast light on shrouded perceptions. Some of these "aha!" moments turned into catchy one-liners, which we compiled into a list; we think they encapsulate the essence of this book.

This list—our Communication Manifesto for Science, if you will—serves as a more philosophical counterpart to the Crib Sheet at the beginning of the book.

1. **Communication is at the heart of any human activity.**
2. **Always ensure that the basic concepts you use are clearly defined.**
3. **Remember: "I communicate, therefore I take action!"**
4. **Differing mindsets can significantly hinder mutual understanding.**
5. **Never let yourself be inhibited by the imperfection of early versions.**
6. **Formal discipline in communication leads to clarity—which in turn leads to prompt execution and minimizes misunderstandings.**

Further Reading

Our Top Ten List

Here follows a (non-ranked) list of ten useful handbooks, along with brief indicators of why you should have them on hand at your workplace—either as physical books or downloaded e-books; some of them may also be accessible through your university library.

- For rhetoric in general: *Rhetoric, A Very Short Introduction* by Richard Toye, OUP Oxford 2013.
- For general aspects of the communication of science: *Effective Science Communication* (Second Edition) by Sam Illingworth and Sam Allen, IOP Publishing 2016.
- For outreach: *The Science of Communicating Science: The ultimate guide* by Craig Cormick, Csiro Publishing 2019.
- For basic scientific peer-to-peer communication: *Eloquent Science: A practical guide to becoming a better writer, speaker, and atmospheric scientist* by David M. Schultz, Springer Science & Business Media 2013.
- For general, visually based scientific communication: *Designing science presentations: A visual guide to figures, papers, slides, posters, and more* by Matt Carter, Academic Press 2012.
- For poster design: *Better posters: plan, design and present an academic poster* by Zen Faulkes, Pelagic Publishing Ltd 2021.
- For design in general: *Universal principles of design, revised and updated: 125 ways to enhance usability, influence perception, increase appeal, make better design decisions, and teach through design* by William Lidwell, Kritina Holden, and Jill Butler, Rockport Pub 2010.
- For graphical design: *The non-designer's design book: Design and typographic principles for the visual novice* by Robin Williams, Pearson Education 2015.
- For intercultural communication: *The culture map: Breaking through the invisible boundaries of global business* by Erin Meyer, Public Affairs 2014.
- For podcasting and storytelling: *Out on the wire: The storytelling secrets of the new masters of radio* by Jessica Abel, Crown 2015.

GPSR Compliance

The European Union's (EU) General Product Safety Regulation (GPSR) is a set of rules that requires consumer products to be safe and our obligations to ensure this.

If you have any concerns about our products, you can contact us on

ProductSafety@springernature.com

In case Publisher is established outside the EU, the EU authorized representative is:

Springer Nature Customer Service Center GmbH
Europaplatz 3
69115 Heidelberg, Germany

www.ingramcontent.com/pod-product-compliance
Lightning Source LLC
LaVergne TN
LVHW010959250326
834688LV00003B/28